DUE FO

QC 24.5 ROT

A PHYSICIST ON MADISON AVENUE

Also by Tony Rothman

Censored Tales (U.K.)

Science à la Mode

Frontiers of Modern Physics

The World Is Round

A PHYSICIST ON MADISON AVENUE

Tony Rothman

PRINCETON UNIVERSITY PRESS PRINCETON, NEW JERSEY

Publication History

Parts of "Instruments of the Future, Traditions of the Past" appeared under the title "Fiddling with the Future" in *Discover*, September 1987.
"The Seven Arrows of Time" appeared in *Discover*, February 1987.
A somewhat shorter version of "The Measure of All Things" appeared under the title "What You See Is What You Beget" in *Discover*, May 1987.
"On That Day, When the Earth Is Dissolved in Positrons . . ." appeared under the title "This Is the Way the World Ends" in *Discover*, July 1987.
"Stranger Than Fiction" appeared in *Analog*, December 1986.
A slightly shorter version of "The Ultimate Collider" appeared in the "Amateur Scientist" column of *Scientific American*, April 1989, under the title "How to Build a Planck Mass Accelerator in Your Solar System." Copyright © 1989 by Scientific American, Inc.
All rights reserved.

Figure 7.2(c) is reprinted by permission. Copyright © 1985 by American Greetings Corporation.

Copyright © 1991 by Princeton University Press
Published by Princeton University Press, 41 William Street, Princeton, New Jersey 08540
In the United Kingdom: Princeton University Press, Oxford
All Rights Reserved

Library of Congress Cataloging-in-Publication Data

Rothman, Tony.
A physicist on Madison Avenue / Tony Rothman.
p. cm.
Includes index.
ISBN 0-691-08731-8
1. Science—Popular works. Cosmology—Popular works.
3. Physics—Popular works. I. Title.
Q162.R82 1991 500—dc20 90-44941

This book has been composed in Linotron Palatino

Princeton University Press books are printed on acid-free paper, and meet the guidelines for permanence and durability of the Committee on Production Guidelines for Book Longevity of the Council on Library Resources

Printed in the United States of America by Princeton University Press, Princeton, New Jersey

1 3 5 7 9 10 8 6 4 2

Contents

Preface

(On Eyeing The *New York Times* Offices from 43rd Street)

If you should happen to be a guest at a European cocktail party, take note of the first question put to you. You will find that small talk sometimes begins with art, frequently literature, occasionally philosophy, these days of course politics, food shortages, corruption (they are the same). Next time you wander into a cocktail party on this side of the Atlantic, try the same experiment. You will find that the opening question is invariably "What do you do?" or words to that effect. It is an innocent enough question, brief, to the point, befitting a people who have for two centuries been admonished to rise early.

But the expected nature of the reply is coiled within the question itself. You see, America, being a practical nation, is also the home of the short story and the multiple-choice questionnaire, lately the sound bite (my six seconds are up, sorry). To answer that you are, for instance, both a scientist and a writer is to darken two of those little ovals with your No. 2 pencil, an error you are sternly warned to avoid before the standardized exam, and the Educational Testing Service computers have never been programmed for such a response. Neither, evidently, have cocktail-party guests. After an awkward exchange full of puzzled "do you mean . . . ?" 's and apologetic "not exactly" 's, you find yourself ignobly relegated to the status of the basilisk, one of those mythological creatures defying zoological classification. On the one hand, any scientists present begin to suspect with covert glances that you are not entirely for the cause. On the other hand, writers (in particular journalists) assume that because you are a scientist you have no business putting pen to paper (another expression doomed to follow "dialing a phone number" into anachronism). The scientist-writer is accused of having an identity crisis, and discussions ensue over whether the proper title is scientist-writer, scientist/writer, or scientist and writer.

I blame society. Seriously, the identity crisis seems to me largely of American manufacture. In Europe there exists a solid tradition of intellectuals who participate in life. Authors with technical and academic credentials have always been prominent in literature. Chekhov, Zamyatin, Bulgakov, Eco, Bernal, Snow, and Dyson come immediately to mind. Czechoslovakia has a playwright for a president, Lithuania a music professor; Helmut Schmidt, former chancellor of Germany, has recorded the Mozart Triple Piano Concerto. The Soviet Union publishes approximately six times as many popular science magazines as the United States, and virtually every leading Soviet scientist writes for them—it is considered part of the job. I have often wondered whether it is coincidental that Russian lacks a word for scientist; you say *uchyoni*, "scholar."

From my point of view, I have always just been doing the job: researching, writing, attempting to reach classrooms farther removed than the walls of my immediate institution. For that reason I prefer the connotations of "scholar" to those of "scientist," "academic," and, in particular, "journalist." My unease with this last designation explains, to a large extent, the slightly peculiar nature of *A Physicist on Madison Avenue*.

You see, nature abhors a vacuum. Since the great majority of scientists do not write, since the majority of writers know nothing of science, and since a numerical code for scientist-writer (scientist/writer?) does not exist on the Internal Revenue Service's Schedule C, a new profession has come into being: the science journalist. It is encouraging that someone, somewhere, has begun to recognize the importance of science for society. I am happy for journalists to do a job most scientists won't (communicate), and during my year on Madison Avenue, which surfaces in this collection, I not only became friendly with many members of the journalistic species but, to my everlasting surprise, became classified as one.

Here we come to the heart of my dilemma. We have indeed entered an age when to report the news is to become the news, when journalists are increasingly associated with (and sometimes claim proprietary rights to) the events they cover. These days, when I discuss proposals with editors I am occasionally asked what science journalism credentials I carry (none). More and more frequently, editors request that I write from the perspective of a journalist, not a scientist. Several years ago I received an unsolicited application for the Knight Science Journalism Fellowship. Never having heard of the program, I made inquiries, to be told I was probably ineligible because the grant was for journalists who write about science, not for scientists who write in journals. In a word, "science journalist" has phased out the always-problematic scientist/writer (scientist and writer?); it is now the box that can be filled in at cocktail parties and probably soon on the IRS's Schedule C.

My resistance to the classification surfaces in *A Physicist on Madison Avenue*, which includes a few of my excursions into the world of science journalism. As the title implies, I consider myself a throwback to that previous generation of scientists who write; yet—for reasons that will become self-evident—this particular throwback often finds himself on more journalistic territory.

The four *Discover* articles, which formed the starting point of this collection, are good illustrations of my "schizophrenia." When *Discover* commissioned the pieces, the editors there expected me to write up certain developments in my field. I was, as I understood it, to take on the guise of a journalist and report the theories and thoughts of other researchers. In principle I was ready to accommodate them. Most mortals have few enough ideas, and the individual's contribution to the whole is relatively trivial. Yet, what happened with the *Discover* articles happens with most of my journalistic pieces: my upbringing as a physicist proves impossible to disavow. I begin to interview other researchers, but within a few minutes it becomes evident that we speak (more or less) the same language, and the distinction between interviewer and interviewee breaks down. We soon end up arguing like colleagues: my subject explains his theories in full technical regalia; I ask annoying questions and try to find flaws in the arguments; on one or two occasions I have ended up suggesting possible research projects. To make matters worse, in the areas covered by the *Discover* articles I have actually done research, sometimes more, sometimes less.

The result of all this is a series of articles lying on the borderline between conventional journalism and personal essays. Sometimes, as in the case of the *Discover* articles, I try to keep my own opinions to a palatable level. (Nevertheless, my editor there remarked that I like to "write from a position of authority"; he invests me with more ambition than I possess.) In other cases, such as the title essay, the outcome is as much a personal memoir as a science article.

To me, the approach has never needed justification. The popularization of science is not new, and after all, I am not the only scientist who writes for an audience wider than the 1.7 readers who read the average scientific paper. Asimov, Dyson, Gould, Lightman, Sagan, and Trefil would also not respond to the call of journalist. Given, however, that pure researchers suspect scientists who write of treason and editors now expect you to have a *New York Times* column before granting you an ear, let me outline a few reasons why scientists should not abandon the popularization of science to journalists.

The first point is that a scientist's training gives him or her a level of expertise few journalists enjoy, and this should manifest itself in the accuracy and depth of the coverage. When writing about a familiar field a scientist does not need to learn the terminology (a large part of the battle); it is already under the belt. And when an idea is understood, it can be written up in fewer words than when it is not understood. At the risk of offending my colleagues at the *Times*, many of their *Science Times* articles are twice as long as they should be. I can be more specific.

The second point is that most scientists reason alike. They are particularly sensitive to the internal logic of an argument and have been trained to point out contradictions. They are equally attentive to the basic assumptions on which an argument rests and whether they have any connection to reality. By examining assumptions and thenceforth following the path of logic, a physicist still can find the correct questions to ask of a biologist (and vice versa), once the terms have been demystified.

The third point is that a scientist, by virtue of experience, builds up a healthy sense of discrimination. (I suspect this is in part because scientists change jobs less frequently than journalists.) There is a discrimination in when a number means something and when it doesn't. During the editing of the *Discover* pieces I found myself in pleasant, although never-ending, discussions with the famed Time-Life fact-checkers over the age of the universe. I routinely wrote 10 billion years, because the truth is we do not know the exact age and 10 billion is almost certainly correct to within a factor of two. The fact-checkers, not being scientists, insisted on 15 billion years because that number had been published in one of *Discover*'s past issues and they did not want readers to become confused. Similar disputes took place over whether the average density of matter in interstellar space is one atom per cubic centimeter or one-half. Yet other numerological debates will be found in the title essay of this volume.

Discrimination also extends to a sense of when an idea is promising and when it isn't. I happened to be on the Editorial Board of *Scientific American* when the cold fusion story broke in 1988. The *Science and Citizen* editor asked for my advice. I told him in my usual diplomatic manner to ignore it all altogether, that by the time our next issue hit the stands the story would have disappeared. From a journalist's point of view, he could not follow my advice (and did not), but by the time the next issue hit the stands, cold fusion had all but evaporated, and we were forced to run several follow-up articles explaining why. At the risk of offending my colleagues at the *Times* again, if I read about a development in its pages before I read about it in a scientific journal or hear about it from a scientist, my rule of thumb is to discount it.

These arguments are, admittedly, a little lopsided; they attempt to show how scientists can benefit the practice of popularization. However, popularization can also benefit a scientist's research. Andrei Sakharov remarks in his memoirs that many of his scientific ideas came to him when he was writing popular articles. (I should say a Russian popular article might be called "semitechnical" in this country.) Sakharov points to a common phenomenon: one tends to get ideas while *not* engaged in research. Von Weizsacker got his ideas on galaxy formation, I am told, while observing the motion of cream in a cup of coffee. Dyson claims he ran across the idea for the "Dyson sphere" in *Starmaker*, a science-fiction novel by Olaf Stapledon. On a somewhat less grandiose level, several of my own scientific papers have resulted from popular articles. "The Epoch of Observational Cosmology" (Chapter 6 of this volume) is a case in point. Although the technical version was published first, it resulted from my meditations about how a civilization in the far future would view the cosmos. I believe that writing popular articles frees up the imagination from technical details and allows new ideas to take root. Thus I recommend the practice to my colleagues.

There are more general reasons why it is important for scientists to write for a wide audience. I began this essay remarking that the scientist/writer split seems to be an American phenomenon. If so, it is not merely the result of increasing specialization but a part of the general divorce of science from the rest of society. America has always been inclined toward practical pursuits, and the inclination appears to have accelerated since the Apollo 11 moon landing, which was forgotten within a year. The decline of real funds for basic research by thirty percent in the past five years is a serious manifestation of the trend. An amusing manifestation surrounded "On That Day, When the Earth Is Dissolved in Positrons," which concerns the end of the universe. It originally appeared in *Discover*, and as a result of the striking black cover the staff had designed, I ended up giving a number of radio interviews (and almost became a guest on the Oprah Winfrey show). Every interviewer asked me the same question: "Why do you study this stuff?" Quickly I got wise and when the last interviewer fired his assault, I retorted, "Why did you call me?" "Well, because it's interesting." "There you go."

I don't know if Oprah would have been satisfied with the response. Although I don't own a television, I have the impression that scientists are not frequent guests on her show. Neither do I have the impression that athletes are asked why they work out, artists why they paint, or poets why they scribble. The singular position of scientists is "enhanced" by filmmakers, who have increased the distance between the scholar and everyone else. If I can encroach on the territory of "A Physicist on Madison Avenue" for a moment, Hollywood, with a consistency that is truly remarkable, relegates scientists to the roles of nerds and madmen. The days when Greer Garson discovered radium or Paul Muni developed magic bullets seem to be gone forever. Scientists do enjoy playing the role of Faust, the scholar in eternal pursuit of Truth, far removed from the concerns of humankind. But even Faust, in Goethe's rarely read Part II, meets his maker at a public-works project. And one has only to list the world's leading political dissidents of a few years back to realize they are practically all physicists.

How is it, then, that the Hollywood portrait is so seriously at odds with my own? I suspect the mismatch arises, in part, because most film directors ended their academic careers as undergraduates, when they had ample opportunity to observe laboratory rats and computer hackers at close range. But having left the hallowed halls at an early age, what they fail to realize is that these people are generally not the ones who make it to a Ph.D., certainly not past it. Scientific research takes character. Most of the scientists I know—and most of the people I know are scientists—are reasonably normal human beings with families and hobbies. They invariably know more about the rest of the world than the rest of the world knows about them. Some jog.

Yet, if only a dozen scientists in the country expose themselves to the world at large, Hollywood will persist in tradition. Someday I would like to see a television series with a scientist as a leading character, a scientist who also happens to be a human being.

It is unlikely that *A Physicist on Madison Avenue* will rectify the situation. These are, after all, basically just popular science essays about subjects that interest me. Most, being of a journalistic nature, have not been written from the skeptical point of view taken in my previous *Science à la Mode*. Nevertheless, as I have said, the present essays have all been written from the point of view of a physicist (of which skepticism forms a large part), and I hope, aside from being entertaining, they give some clue as to how a physicist thinks. The first two essays in particular show how a physicist (or physicists) might tackle problems arising on foreign territory: Madison and other Avenues. The remaining venues are closer to the concerns of a cosmologist: time, the galaxy, the universe.

Because the core essays in this volume were written for high-circulation magazines, they are shorter and easier than those of the previous collection. Those who found the *Science à la Mode* essays too difficult will probably be happier here. Those who found them too vulgar should probably skip *A Physicist on Madison Avenue* altogether.

Because several of the pieces deal with cosmology, there is a bit of repetition when dealing with important themes: cosmic nucleosynthesis, in particular, and inflation. I have decided to let the repetition stand so that the essays may be read separately, although there is a case to be made for elimination in the cause of unity. Still, as they say, repetition is the mother of learning.

Four of the essays in *A Physicist on Madison Avenue* were written for *Discover* magazine when it was owned by Time-Life Incorporated. Editors often like to remain anonymous. At the risk of making them blush, I would like now to thank the entire *Discover* staff, in particular Gil Rogin, the Managing Editor, Peter Carry, the Executive Editor, and Kevin McKean, my own senior editor. I can safely say I have not encountered a more professional team of journalists. I particularly enjoyed the no-holds-barred approach and the freedom they gave me to write on subjects of interest to me. I am also honored that they were sufficiently pleased with my work to put every story on the cover. My only regret is that Time Inc. sold the magazine.

For this volume I have gone back and begun revisions from my original versions, picking and choosing improvements from any of the subsequent variants. I do not intend to slight the work of the editors, in particular Kevin, whose suggestions I have often incorporated here. The opportunity to publish the essays in book form has given me much more space than was available in the cramped pages of *Discover* and has allowed me to restore cut material. And *Discover*, like all major magazines, is highly copyedited. Approximately ten editors have a shot at each article. So long as *A Physicist on Madison Avenue* is to be my book, it may as well be in my voice.

Two of the *Discover* pieces, "The Seven Arrows of Time" and "On that Day, When the Earth Is Dissolved in Positrons. . ." appear here not significantly altered from their magazine appearance. "The Measure of All Things" is expanded somewhat more. "Instruments of the Future, Traditions of the Past," intended as a grand overview of scientific instrument making, unfortunately got caught in the sale of the magazine and appeared under the new management, which was forced to cut it by seventy percent. Its appearance here may be considered its first publication and is very much a physicist's view of instrument making. Journalism students may find it educational to compare what physicists consider interesting to what editors consider interesting. Followers of musical acoustics may note the absence of a direct interview with Arthur Benade, probably the country's leading authority on the physics of woodwinds. At the time I was writing the article, Benade was terminally ill with cancer; I was grateful that he and his wife nevertheless showed sufficient interest in the project to put me in the care of Benade's postdoc, Peter Hoekje.

"The Epoch of Observational Cosmology" and "Alternative Cosmologies" provide good companion pieces for the *Discover* articles. The former was written as such; the latter was written for *Scientific American*, but after I resigned it was never published. For "Alternative Cosmologies" I am joined by two frequent and distinguished collaborators, Richard Matzner, Director of the Center for Relativity at the University of Texas, Austin, and G.F.R. Ellis of the Department of Applied Mathematics at the University of Cape Town. "Alternative Cosmologies" covers much the same ground as "The Garden of Cosmological Delights" did in *Science à la Mode*, but, because it was intended for *Scientific American*, the level is somewhat higher. It is the most difficult article of the collection. Nevertheless, it seems to me that the preceding articles give ample groundwork for it, and since "Alternative Cosmologies" is so closely connected to them, I have included it.

The last two articles retreat from cosmology to the slightly smaller scale of astrophysics. My collaborator for "Stranger Than Fiction" is David Aschman, a particle physicist at the University of Cape Town, who brought to my attention the mystery of Cygnus X-3. Once again, a comparison of our account of Cygnus X-3 with those that appeared in newspapers and the more recent newspaper reports of the similar phenomenon involving Hercules X-1 might be instructive to journalism students. The last piece, "The Ultimate Collider," appeared as the 1989 April Fools' joke in *Scientific American*'s Amateur Scientist column under the pseudonym Antoni Akahito, Anthony Rothman in Japanese. (The Japanese are most likely to build the thing.) I originally wrote it hoping to prove that a Planck-mass accelerator would be impossible to construct. I finished it being less sure. It may say less about accelerators than about a physicist at play, which, I suppose, is another way of saying what *A Physicist on Madison Avenue* is all about.

Princeton
March 1990

A PHYSICIST ON MADISON AVENUE

CHAPTER 1

A Physicist on Madison Avenue

I

Life has a stubborn tendency not to follow the course you've charted for it. My own trajectory, painstakingly calculated to place me in an academic orbit, went awry early. In late 1986 I washed up on the shores of the New World, having survived two years of demonstrations, riots, teargas, and cosmology at the University of Cape Town, South Africa. I was penniless, abandoned by every funding organization known to man. So I set off on the uncertain life of a freelancer. For once luck was with me, though, and after four lean months I was writing for *Discover* magazine, sanguine of the future. But Fortune again intervened, *Discover* was sold, and my financial situation quickly reverted to its previous state. The threat of immediate starvation exacted from me a drastic step: I accepted an offer for an editorial post at *Scientific American*. And thus I found myself, unwittingly, a physicist on Madison Avenue. . . .

II

Madison Avenue. The real world. I had heard of such a place ever since my undergraduate days, but never expected to set foot in it. As far as I could make out, the real world was a land of three-piece suits, high finance, corporate mergers, and power lunches. Inhabitants of the real world went to discos, watched "Dallas" on TV, and spent their vacations in Disneyland. At least, this is what I have read about the real world in tabloids.

The contrast to life in academia could hardly be greater. I do not need to describe the existence of a university scientist in great detail, because you are already familiar with it. The makers of *Revenge of the Nerds* know, as do the millions who have seen it, that all scientists when young are undernourished sociophobes who relate best to a computer terminal through coke-bottle eyeglasses after midnight in a basement laboratory. As they age they put their phobias behind them to become positively pathological, with only one thing on their minds: dissection of the nearest alien. (*ET*, *Splash*, *Lord Greystoke*, and *Iceman* provide irrefutable evidence for skeptics.) If the nerd is lucky enough to avoid seduction by the Dark Side, he gradually sheds his cocoon to emerge in the shining new guise of—a dork! a term Murphy Brown, having been afflicted by one in a recent episode, can probably explain as well as I.

And so I arrived on Madison Avenue armed with the knowledge that I was at best useless and at worst dangerous. That I was at least useless was brought home to me after five minutes on the job when I asked one of the stockroom clerks for a blackboard. The fellow was as shocked at the request as I was at the reply: "Impossible."

From my point of view it was natural to think that a blackboard might come in handy. The world of a physicist consists of natural laws—such as conservation of energy and momentum—which govern the behavior of everything in the universe and which cannot be broken by any mortal. The world of a physicist consists of mathematical derivations and proofs, which go into the logical construction of a theory. And the world of a physicist consists of experimental evidence by which theories are tested, for without experimental evidence all theories are equally good—or equally bad.

The idea that the world is lawful as opposed to lawless is fundamental to a scientist's outlook. The most important message of the natural laws is that some things are impossible (like getting a blackboard). The principle that some things are "no go" leads in turn to the sobering realization that none of us are granted absolute control over the world. Because the laws determine the rules, scientists learn the habit of following the evidence where it leads, regardless of whether the eventual conclusion is compatible with Marxist philosophy or Reaganomics. This is called scientific method. Such ideas, with their imposition of constraints, are not popular in America, where to "take control of your own life" has become something of a national pastime. Such ideas run counter to everything *Self* magazine stands for.

Nevertheless, because Madison Avenue was part of the real world, I had assumed it would obey the same laws of physics as everything else. Here I was mistaken. Madison Avenue, unlike the world of a physicist, consists of advertising departments, circulation departments, promotion departments production departments, and editorial departments (in order of decreasing significance). Advertisers and circulation managers use terms like "base rate," "draw," "sell-through," and "volatility." "Energy" and "momentum" do not figure prominently.

The absence of conservation of energy in a department devoted to filling the pages of *Scientific American* with liquor ads might not surprise inhabitants of the real world. It did me. The reality survey shows why.

III

Not long after my arrival Jonathan Piel, the Editor-in-Chief (or Editor, to use the formal *Scientific American* term), told me that the majority of the readers were not scientists. The news disturbed me greatly because since the days when I actually read the magazine I had assumed that most of its readers were students, academics, or researchers who wished to learn something outside their fields. No, I was assured that the readers were by and large managers, members of the business community, and "busy executives."

The identity of the readership is not an academic question. Management has hoped to launch two new publications, *Trends in Computing* and *Scientific European*, which are geared more toward business and industry than to the scientific community. The parent magazine has made similar excursions into reality. In mid-1988 it inaugurated a new department, *Science and Business*, and 18 months later expanded it from three pages to ten; the traditional *Science and Citizen* section—which consists of short news items—is also now twice as large as it was several years ago, presumably on the theory that executives don't have time to read anything longer than the average article in *People* magazine.

My first thought was that the readership question should be simple to answer: make up a subscriber survey and ask the readers who they are. I discovered I was not the first person to have such an idea. But the survey I got my hands on six months later was not written by a simpleton; it was written by a market research bureau, and market research bureaus do not want to find out who the readers are (which is irrelevant) but how rich they are (which determines whether they can afford a Mercedes).

The survey contains precise questions about income level, length of time an average issue is kept, and favorite departments, but anything so straightforward as "Where do you work?" or "Are you a scientist?" is absent. Instead, the market research bureau sorts all readers into categories under the heading "type of business." Included are such real-world occupations as real estate, insurance, government, engineering, education services health services, and "other." The reader is also given a choice of real-world job titles: Chairman of the

Board, Owner, Partner, Vice President, etc., CEO, General Manager, Professor/Teacher/Other Educator, M.D./D.O./D.D.S., and "Other."

The first thing you notice is that if you do not work in a business you are assumed not to exist. Students and scientists (along with the dodo) are denied ecological niches; the closest slot is "education services." The same pattern follows for the job titles. "Professor/Teacher/Educator" is the only pigeonhole for an academic or researcher, while there are no less than six possible slots for executives and managers.

Still, the psalmist has said, "Knock and it shall be opened to you." In this survey those with executive or managerial titles accounted for 27 percent. Twenty-seven percent is sizable—it comprises the largest known category. But those answering the description Professor/Teacher/Educator made up about 24 percent, and "other" 26 percent. Depending on your inclinations you might venture to guess that the 50 percent consisting of educators plus "others" are connected in some way with academia or research.

Inclinations have a subversive tendency to be confirmed. Adding up all the responses to the "type of business" question, you find they total only 82 percent—18 percent of the readers vanished from the real world altogether! About 27 percent managed to locate themselves in education services and 11 percent in "other." A missing persons bureau might be set up to find out who all these people are. At the lower end, the number of academics who slipped through the cracks could be 29 percent; at the higher end, 56 percent. A physicist would say that these numbers are consistent with the previous guess.

Is it a coincidence? I think not. Another question on the survey asks for educational level. Eighty-five percent of *Scientific American* readers respond that they have finished college. About 22 percent claim to hold master's degrees, and an additional 40 percent hold doctorates. I personally know no busy executive with a Ph.D.

One might suggest that the management has been pushing the magazine toward business on the basis of predisposition—not on the basis of evidence. One might still want to ask what a subscriber survey has to do with conservation of energy. If either thought has occurred to you, you may be less an inhabitant of the real world than you think.

I said earlier that the habit of obeying natural laws forces a scientist to follow the evidence where it leads. The *Scientific American* survey was at best not designed to find out who the readers were; at worst it was designed not to find out who the readers were. In either case the design makes it easy to ignore the truth. Many firms hire "answer analysts," people who look for data to support the company's position on a given issue. You want to prove that acid rain is not caused by coal-fired power plants? We'll find you the data. You want to prove that most *Scientific American* readers are businessmen? I'll design the questionnaire. The answer-analyst approach runs exactly contrary to the scientific spirit, which is to go out, collect data, and *then* decide on a conclusion, not vice versa.

What of my own bias that *Scientific American* readers are predominantly linked to academia? We all have preconceived notions. A good scientist knows when to throw them out. He cuts the data as many ways as he can to check his answers (I gave you three cuts above), and if the data are not sufficient, he goes out and collects more. I am willing to wager that a reasonable survey will support my conclusions and not those of the marketing bureau. What is a reasonable survey? Make up the job categories *after* you have asked the readers what they do, not before. Then if the data prove me wrong, I'm wrong.

Underlying this entire exercise has been a hidden factor: I have been banking on my number sense. If devotion to Law is the primary characteristic that separates physicists from the inhabitants of the real world, then not far behind is the physicist's peculiar relationship with numbers. I am not speaking of a facility for arithmetic. Most physicists are probably not as good at arithmetic as accountants or bank tellers. But what a physicist develops during his training is a sense of when an answer is right and when an answer is wrong—in other words, when a number means something and when it doesn't. Unfortunately, a number sense is itself enough to cancel a physicist's visa to the real world, as I discovered during the great statistics battle I waged against the full might of *Scientific American*'s circulation department.

IV

In October 1988, after I had been on the job about five months, Harry Myers, the President of the Magazine Division and Publisher, asked the members of the editorial board to try to discover why certain issues of *Scientific American* sold better on the newsstand than others.

You must understand that some 80 percent of the magazine's sales are in subscriptions but each month about 100,000 people across the country stop by the local newsstand, gawk at *Penthouse*, and buy *Scientific American*. From month to month the numbers vary; in the last few years the average has been 96,000 but sales range from roughly 70,000 to 120,000. Anyone looking at month-to-month figures would see no obvious pattern; it was just this that troubled Harry Myers and members of the circulation department. In particular, Myers wanted to know why the February 1988 issue sold an extraordinary 122,000 copies. Was there any way to stop such fluctuations and keep sales consistently on the high end? I should say that "fluctuation" is the term of a physicist. In the real world the presence of fluctuations is referred to as volatility. I do not like the term, with its implications of containable temper outbursts, but "volatility" will continue to be used on Madison Avenue and advertising executives will continue to believe that everything is under their control.

What causes sales to fluctuate? In the real world, cover design has a strong influence on sales. Bill Yokel, the Circulation Manager emeritus, has long held that predominantly red covers sell well and blue covers sell poorly. Another idea, attributed to Harry Myers, is that the more abstract the cover the better. Because in the last few years the President of the Magazine Division and Publisher, not the art director, has had final approval of the cover, the idea resulted in (among other things) a series of abstract covers during the summer and early fall of 1988: a gravitational lens that looked like a blob, an optical illusion that looked like several blobs, a computer-generated crystal that looked like it came out of the final sequence of *2001*. Still, doctors recommend.

Accepted wisdom also has it that animal covers are bad bets. Regular readers of the magazine may recall that the only animal covers in the last several years were the cheetah on May 1986 and the panda on November 1987. They sold so poorly that animals were banned from the cover once and for all. Yet another piece of conventional wisdom is that there is a seasonal variation in sales. I can't remember every month that does better or worse than average; ask your nearest circulation manager for the exciting details.

When the volatility question was officially announced, each editor was given the 1987–1988 sales figures and photocopies of all the covers and table-of-contents pages, and after several weeks we convened an editorial meeting to discuss the matter. The covers for the two years were posted on the wall. A dozen editors sat down around the executive table and made themselves comfortable; the debate began.

Everyone present had one theory, if not two. Aside from the blue, red, and abstract cover hypotheses, there was the "strong, clean design" hypothesis, which I have never entirely understood. Its advocate maintained that "the best covers sold the best and the ugliest covers sold the worst." Jonathan Piel proposed that "man-and-machine" covers were the most popular. Yet another editor made the novel suggestion that contents were more important than cover: every issue with a major policy article or a math article or a computer article sold better than average. She bolstered her case by pointing out that at most newsstands the cover of *Scientific American* is buried behind *Popular Mechanics*.

The discussion went on for one and a half hours. As the other editors argued, I sat there morosely trying to imagine a similar gathering at the editorial offices of *Cosmopolitan*, where all covers to emerge are essentially identical. "Every centimeter of cleavage sells 90,000 copies. Blue mascara sells better than brown. Sex after marriage in the upper left-hand corner does not do as well as how to carry on an extramarital affair." I couldn't buy it.

After sufficient grumbling I was given the floor. It would be impressive to say I was dueling off the cuff, but I am not so brave and had come prepared. When the sales figures were given to me I did the first thing any physicist (probably even Feynman) would do: make a histogram that showed the number of issues selling a given number of copies. Why?

If you don't want to call it instinct, call it training. It seemed obvious that the factors influencing the sales of a magazine are many: the weather, the cover, the contents, the mood of the buyer, the budget deficit, a chance romantic encounter at the newsstand, perhaps the phases of the moon. In fact there must be so many factors influencing sales that it becomes impossible to isolate any one of them, and the result is essentially *random*.

For this reason I knew that even if some slight pattern existed, it would be buried in statistical noise and extremely difficult to detect. On the other hand, if the newsstand sales were experiencing genuinely random fluctuations, that should be pretty easy to show by a histogram. Physicists are unscrupulous opportunists, so I opted for the easy way out. Within about 15 minutes I had my answer: the 22 issues for which I had data formed a Gaussian distribution, at least as close to a Gaussian distribution as you could hope for with only 22 points.

You may not have heard the term Gaussian distribution, but you have heard the terms normal distribution or bell-shaped curve. They are synonyms. Many random processes result in Gaussian distributions, so many that the bell-shaped curve has become the symbol of randomness.

That the sales fell into a bell curve was not an entirely foregone conclusion. I also plotted the sales of the international edition, and try as I might, I could not fit them into a Gaussian. And non-Gaussian it has remained to this day, even as more data has become available. Therefore something nonrandom is evidently taking place. From the shape of the distribution I might conjecture that the international edition is largely selling out, but I am reluctant to frame hypotheses.

Sane readers may skip this paragraph. To conspiracy theorists (eternally insistent on patterns) I admit that if sales were entirely determined by a single X-factor, say the amount of purple on the cover, and *if* the amount of purple were fluctuating randomly, sales would then fall into a normal distribution. In this case the sales would be exactly *correlated* with the amount of purple despite the random fluctuations. Such a scenario, while logically possible, is beyond far-fetched. Even if such an X-factor existed, its random behavior would make spotting it difficult; if two or three random X-factors were driving sales, to sort out the correlations would be, as I've said, hopeless. Simply defining possible X-factors is an exercise in frustration. "Amount of purple" may be a good variable because it is easy to quantify (80%, 93%, . . .). But the editorial suggestions "man-and-machine," "animals," and "abstract" by their very nature cannot be random variables. Sane readers may continue.

Although I was doubtful of the outcome, to check some possible correlations was easy enough. When I sat down with the sales figures I made not only a histogram but a chart that divided covers into various categories: predominantly blue, predominantly red, high-tech, artistic, nature, abstract, animal, and so on. Admittedly the categories were subjective, but, I thought, not unreasonable. I took the two years' worth of data and put each issue into the appropriate bin, marking whether it had sold better than average or worse than average.

I could see no correlations—roughly half the issues in each category sold better than average and half sold worse than average. Actually I'm lying. Two patterns did emerge: 3 out of 4 "artistic" covers sold worse than average, and, to my surprise, red covers sold better than blue covers. Of the 5 red covers, 3 sold better than average, and of the 9 blue covers, 6 sold worse than average. Had I confirmed Bill Yokel's theory? I will explain later why I didn't take this very seriously. The lack of a convincing pattern in my chart contributed to my morose silence during the editorial meeting—despite continued commentary on "bold, striking designs" and "man-and-machines." Perhaps I needed a shot of Geritol.

None of my results could exactly be called earthshaking. What stunned me was that apparently no one on Madison Avenue had ever made a histogram. I soon learned that a histogram is not an obvious thing to draw. For the meeting, Ed Bell, one of the art directors, had plotted the sales figures for 1987 and 1988 month by month. From Ed's point of view a month-to-month graph was a natural way to look at the world. But it wasn't helpful: the graph zigzagged up and down in no recognizable pattern. In fact 1988 zigged largely where 1987 zagged, canceling out any possible trend and allowing you to conclude absolutely nothing.

So I stood up. Since blackboards were still lacking, I had come dressed incognito with a flip chart. "The very number of hypotheses I've heard," I began, "convinces me that my hypothesis is correct—there is no pattern." I explained that the sales figures formed a Gaussian distribution with a mean of about 96,000 and a standard deviation of about 13,000. Because the fluctuations were random, about 5 percent of the points should lie farther than two standard deviations from the mean. The February 1988 issue, which sold 122,000, was just such a fluctuation. The only remaining mystery was how the circulation department had provided sales figures for the November and December 1988 issues. November returns had yet to come in, and the December issue had not even appeared. Those points I included out of perversity but marked them "magic."

The presentation appeared to be going smoothly. Midway the Editor turned to Ed Bell and, in a voice perfectly balanced between jest and earnest, quipped, "Institute for Advanced Study school of market analysis." (Later a mathematician friend recommended I let them think that.) The ambiguity of the Editor's remark presaged things to come. After I explained the details of the histogram, a voice was raised.

"You've used absolute sales figures."

"Yes, of course."

"What about sell-through?"

I suppose it was inevitable. Throughout the meeting I had noticed that with the exception of Greg Greenwell (who also has a physics background) and me, everyone defined a good or a bad sale in terms of sell-through, which is Madspeak for percentage sale. Each month a certain number of copies are printed, or "drawn," but only a percentage of them are sold. That percentage is sell-through. A good sell-through is considered something like 48 percent.

My interrogator should not have handicapped himself by asking me the same question the previous day. "What you print is a managerial decision," I answered, "what is bought on the streets is determined by the consumer. The correct figures to use are not the percentage sales but the absolute sales."

At the time I did not know it, but what I had just said was sufficient to cause excommunication from Madison Avenue. According to Madthink it is an incontrovertible fact that if you raise the draw you raise the sales. I have never been able to get a coherent explanation for the phenomenon; if you put 20 copies out on the newsstand and only 10 are selling, it is not obvious that putting out 30 will help matters. I imagine the theory is that if you increase the draw you increase the number of newsstands, so more people encounter the magazine.

In either case, to check a possible relationship between sale and draw was easy enough. I merely graphed the sale versus the draw. If you visualize a shotgun fired at a target you have a good picture of the result. There was no correlation whatsoever. Points were splattered all over the place. I tried to fit a straight line to the data, first unsuccessfully by eye, then by calculator—but my calculator went up in smoke.

Even as I told of my calculator's breakdown, I realized there was no reason to assume that sale and draw should be connected by a straight-line relationship in the first place. If the draw is below demand and you increase the draw, then sales should go up. But as you increase the draw above demand it should have less and less effect on sales, until sales asymptotically approach a limit where the draw has no effect whatsoever. The data didn't do anything like that, either. The data were just a mess.

The problem of basing conclusions on sell-through was particularly amusing (or sad, depending on your feelings toward *in vivo* pharmaceutical testing) when it came to the question of animal covers. When I was making my chart of cover types I noticed one fact that I took very seriously: the panda cover had sold better than average. In the world of statistics the First Commandment is *Thou shalt not do statistics with one data point*. I found it extremely odd that animal covers had been banned on the basis of a single experiment. I found it incomprehensible that they had been banned on the basis of a successful experiment. (The 1986 cheetah, for which I had no sales figures, had yet to enter my thinking.)

More than one person noticed the anomaly at the meeting. "The panda sold slightly better than average: 98,000."

"No," came the reply, "only a 38 percent sell-through."

"But this is nuts," I cried (in a rare outburst). "The only reason the sell-through is so low is because the print run was almost 25 percent higher than usual."

You see, once a year *Scientific American* publishes its famous single-topic issue (perhaps not so famous as *Sports Illustrated*'s swimsuit issue, but both are based on the principle of overexposure.) Although normally it appears in September, from 1985 to 1988 it was released in October. As a rule the single-topic issue sells far better than average; rumor has it that the 1988 issue devoted to AIDS was the best-selling issue in the history of the magazine. It would be difficult to argue that the special issue is a random fluctuation. For that reason I routinely excluded October in the "calculations" I have described; here then is why two years has 22 data points.

So. For the October 1987 issue the draw was raised as usual to accommodate the additional sales expected. For reasons known only to God and the circulation department, the draw was not lowered again for November's hapless panda. The sell-through registered low on the Richter scale simply because the print run was too high.

"The moral I take from this story," I said at seminar's conclusion, "is that we should let the art department decide what to put on the cover and not listen to astrological advice from the advertising department."

"The circulation department."

"The circulation department."

Physicists tend not be diplomats, yet another reason why they should be denied visas to the real world. Those who gather around the conference table at *Scientific American* are a very bright bunch, and they understood each point even as I made it. Still, one or two remained doubtful that draw and sale were uncorrelated and insisted that I make a histogram using sell-through. I agreed. The Editor also asked me to make my presentation to the President of the Magazine Division and Publisher and the circulation department. I agreed to that too.

V

Before I recount the climactic events of this tale, it is worth pausing for a word from our sponsor. The problem I have been discussing is probably the most common faced by a scientist: When do the numbers signify a trend and when do they not? Do the data represent a signal or merely noise? The Cold Fusion Fiasco of 1989 was exactly a case of mistaking statistical fluctuations for a signal, as was the Great Neutrino Scare of 1980, when six research groups simultaneously announced that neutrinos had a rest mass.

Psychology is strenuously at work here (in particular, Result Amplification by Stimulated Emission of Publicity). It is all too easy to see a signal when you want to, especially when the amount of data is small and the ego is large. The traditional solution in science is to go out and collect enough data so that the answer becomes as unmistakable as those golden arches. One might say that more data increases the objectivity of the experiment.

In the real world the solution tends to be the opposite: ignore data you don't like. This approach is undoubtedly the correct one, for the real world becomes a much more subjective and exciting place than the cold, objective world of the physicist. Without astrology, ESP, and Shirley MacLaine's past lives, Waldenbooks might very well go out of business, and conversation at Los Angeles cocktail parties would become impossible. Trenchant subjectivity has made the world livable, but it has also destroyed any social number sense and has made the concept of number meaningless.

I do not mean to exaggerate. Periodically opinion polls on the likelihood of nuclear war appear in *Time* or *Newsweek*. "Do you think that nuclear war in the next 10 years is (a) very likely; (b) somewhat likely; (c) not likely?" There is something reminiscent of Chicken Little here: unless such polls influence strategic policy, the average citizen's opinion on the probability of nuclear war has little if anything to do with the actual probability of nuclear war. The philosophical basis of the surveys is that reality is determined by referendum.

A more extreme reality referendum was a poll taken on the likelihood of nuclear winter. Ninety-two percent of the respondents thought that nuclear winter could take place after a nuclear war. It is intriguing that the laws of physics were not consulted.

The 92 percent are close relatives to the salesman who attempted to convince me to buy an optional three-year service contract for a car (one year after they had been introduced) with the argument that "fifty-eight percent of the customers buy them." Rather than carry a joke past its punch line, I leave it as an exercise for the reader to give three reasons why I told the salesman I was not a member of the Pepsi generation.

The distance from my Nissan dealer to Madison Avenue is less than a finger's walk through the Yellow Pages. When numbers become subjective, sell-through becomes as good a variable as absolute sales. In science, most meaningful numbers are ratios: "A as compared to B." Without the comparison, the quantity under discussion lacks a standard of measurement—a convenient state of affairs in the real world. The *New York Times* recently reported that President Bush "remains highly popular with the electorate" because of his 61 percent approval rating. As one reader pointed out, another way of putting it would be to say that Bush has the third lowest rating in the last 40 years—every new president since Truman, with the exception of Nixon and Reagan, had a higher rating.

The use of sell-through by *Scientific American*'s circulation department is the real-world approach. Although sell-through is the ratio of copies sold to copies drawn, if the draw changes every month, there is no fixed standard of reference. To say a 38 percent sell-through is bad means nothing to a physicist until you fix the denominator. The use of sell-through finds its roots in the subjectivist interpretation of the world.

VI

As I "prepared" for my meeting with the President of the Magazine Division and Publisher, the divergence between the subjectivist and objectivist approaches became ever more apparent. My first task was to get more data. The circulation department obliged and soon we had sales figures for the past five years. On the other hand, a statistical software package did not exist at *Scientific American*, so Greg Greenwell wrote a little program to produce histograms and fit Gaussians. It would have taken me far less time to do it by hand but in the information age graphs produced without MacPaint carry no weight. I offered a wager to the Associate Editor that the resulting curve would be Gaussian, a wager he wisely refused: with five years of sales figures, the histogram was as close to a normal distribution as any data I had ever seen.

To show the critics that there was no relationship between draw and sales, I also made a histogram with sell-through as the variable. As I binned the data my eyes grew wider and wider. What was this? The curve had two humps—nothing like a Gaussian. Did raising the draw really raise the sale? No. I was essentially repeating the panda mistake. If you increase the draw from 100 to 200 copies, yet sell 50 in each case, the sell-through drops from 50 to 25 percent. But changing the draw has not changed the sales; the effect is an illusion due to the fact that the standard of measurement has changed. On the other hand, if increasing the draw to 200 consistently results in a sale of 80, then something real has happened.

I was plotting both illusions and realities, which gave me two humps instead of one. The way to rid yourself of the illusion that you are making more money this year than last is to calculate your salary in constant dollars. I did the same thing: I "normalized" each month's sell-through to a constant draw. In other words, I fixed the denominator. This would erase all illusions, and any effect left over would represent a real connection between draw and sales. But nothing was left over; the curve collapsed into a Gaussian.

At about the same time, I developed a "theory" to explain the lack of connection between draw and sales. A golden rule in physics is that when there is only one important number in a problem, that number has to be the answer. When there are two important numbers, the answer must be some simple function of the two of them. You ask a physicist how long it typically takes for an elementary particle to decay, and he will say 10^{-23} seconds. Why? Because the only relevant time you can imagine is the size of an elementary particle divided by the speed of light: 10^{-23} seconds.

In the newsstand sales problem there were only two significant numbers: the mean and the standard deviation. The mean sales must represent something like the demand, and the standard deviation must represent the average monthly variation in demand. So you would expect that if you were drawing several standard deviations above the mean, any purported influence of draw on sales would have to disappear. *Scientific American* is drawn at well over ten standard deviations above the mean sales. There couldn't be any effect. An obvious experimental test suggested itself: lower the draw.

My theories impressed nobody. Even as he handed over the old sales figures, Bill Yokel explained that in my analysis I must take into account the draw, articles run by other magazines, the cover . . . essentially every phenomenon in the universe. We agreed that it did not make sense to go to all red covers: 50 percent of them would still sell less than average. Yokel was not alone in his skepticism of objectivity. The presentation to the circulation department, scheduled for the next day, was canceled "for anon" by the President of the Magazine Division and Publisher. According to Jonathan Piel, the President was simply too busy and the meeting "might" be rescheduled in the New Year.

The message became louder a few days into 1989, when I received a memo from the Editor. It reiterated his claim that "man-and-his-weird-machine" covers had "extraordinary drawing power" with the public.

Piel had enclosed the sales figures for five man-and-machine covers since 1974. Man-and-machine covers include human-powered hydrofoils, walking machines, and other unlikely combinations of sinew and steel. Of the five covers, all but one—the most recent—had sold above average in absolute numbers. My first question was whether Piel had listed all relevant covers. A little research showed that he had missed at least two—but the sales figures for those also proved to be above average. I was dumbfounded. Six out of seven man-and-machine covers had sold above average. Had Piel discovered the Holy Grail? A tremor passed through my carefully forged principles.

Here was a classic case of mistaking noise for a signal. When I finally got the sales figures from 1984 onward, I discovered two apparently real trends: every December sells better than average, and every May sells worse than average. The trends are not pronounced enough to destroy the Gaussian distributions I've talked about, but in the data on hand there are no exceptions. The Christmas spirit (or airport traffic) probably explains the high December sales, and the low May sales could be associated with the close of the academic year, an admittedly remote possibility since academics do not read the magazine. I see no other evidence for seasonal variation.

What does the Christmas spirit have to do with man-and-machine? It turns out that no fewer than three of the seven covers in question were on December issues! At best nothing can be concluded from them; most likely they sold better than average because December always sells better than average. So, three of the data points had to be discarded, and that left four—three better than average and one worse than average. I told Piel that we should try a man-and-machine cover on an issue that was neither December nor May.

My wish came true in March 1989, when *Scientific American* featured a solar-powered racer with driver on the cover. Yokel had gone out on a limb, predicting in a lengthy memo that the superior cover, contents, and interior illustrations would result in sales of 115,000 copies. I admired his bravery; if it sold that well it would be one of the best-selling issues of the past several years. Although based on a seat-of-the-pants hunch, Yokel's prediction was in the best scientific tradition—easy to prove wrong. And wrong it was. The March issue sold only 89,000 copies, noticeably below average. So, we now had five usable data points for man-and-machine: three sold above average and two below average. I would hesitate to claim "extraordinary drawing power" for man-and-machine.

Perhaps the March issue's poor showing was due to its blue cover. Theoretical physicists are famed for their ability to concoct a theory for any occasion, but compared with circulation managers they are in the minor league. I think you can see why I did not take seriously the fact that in 1987–1988 three out of five red covers sold better than average and six out of nine blue covers sold worse than average: a little more data and everything could change. (It seems almost unsporting to point out that one of the red covers was in December and two of the blue covers were in May. I will be accused of downright protectionism if I remind you that the cheetah was also a May cover.) As for "artistic" covers, they included everything from Renoir to architecture to dyslexia to hunger. A division into "art" and "social problems" would have recategorized the conclusion.

VII

I had almost forgotten about the meeting with the circulation department when suddenly, in mid-January 1989, Jonathan Piel told me it was on for the next day. After assuring him that I wasn't going to say anything I hadn't said before, I dusted off my flip chart and told Greg to prepare his graphics. Thus armed, we prepared to circulate. Apart from Piel and Yokel, Bob Bruno, Director of Circulation, was present, along with John Moeling, the Associate Publisher/Business Manager. Harry Myers, the President of the Magazine Division and Publisher, would "be late."

The meeting began innocently enough. As promised, I revealed no surprises. The audience seemed to buy the Gaussian. When I explained why I thought that sell-through was not an appropriate variable, Bob Bruno nodded. "You know, I've been saying that we should look at absolute sales and not at percent—it's too hard to tell what's going on." The physicist and the Circulation Director agreed. A new age for man was dawning.

I displayed the graph of sale versus draw showing the lack of correlation and gave my interpretation of it. Then I suggested my experiment—lower the draw. And the calm ended. If it were physically possible to recount the objections that bombarded me when I uttered those words, I would do so, if only to heighten the drama, but they were couched in Madspeak and no translator was available. It seems to me that the arguments boiled down to one of psychology: if a news dealer only buys three copies of the magazine and you tell him you are going to take one away, he won't buy any. That may be; I have never understood psychology.

My proposal to save *Scientific American* tens of thousands of dollars each month while furthering the cause of scientific research was thus shot down. However, the gunfire had only begun. Just as I made my appeal to let the art department decide the covers, Harry Myers entered. When he caught sight of the plea, scrawled brazenly in red magic marker across the flip chart, he did not hesitate to take the offensive.

"Who's been putting pressure on the art department?"

I bow to the Editor for his answer. "You and I have."

But the President of the Magazine Division and Publisher evidently hadn't come to listen. He asked me whether I believed that some subjects were naturally more interesting than others, to which I replied yes—to me, but *Scientific American* had half a million other readers. "For every Rothman there's an anti-Rothman," added the Editor. The President of the Magazine Division and Publisher asked whether I believed that some magazines were naturally more interesting than others, to which I replied that wasn't a fair question—I had been asked to look at some numbers, not make value judgments. The response did not exactly pacify him.

"Now wait a minute—you've talked an hour, now I want to ask some questions."

I conceded that a magazine on high-energy astrophysics did not have as much intrinsic appeal as the *National Enquirer*, which led the President of the Magazine Division and Publisher to ask whether a magazine has a "natural" circulation. I said yes—every number I had seen indicated that *Scientific American* had found its level of neutral buoyancy. Despite the new format, the new cover design, and the new departments, the mean sales had not changed by more than a third of a standard deviation in five years; everybody who wanted to buy the magazine was buying it, and if the publishers wanted to increase the circulation radically they should begin producing videos.

The only thing lacking in the discussion was any connection with the previous hour. John Moeling finally interceded: "There's nothing he's said you can really argue with."

Moeling has my everlasting gratitude. The pause in the disaster gave me time to show the President of the Magazine Division and Publisher the five-year Gaussian and explain its significance. "You shouldn't be concentrating on the fluctuations, you should be concentrating on the mean."

For a moment, it seemed as if peace would reign. Then Moeling asked what would happen if we could eliminate the tail end of the Gaussian—those five or six issues that sold the worst.

I made the mistake of answering tautologically: "Then you'd decrease the volatility."

"But that's exactly what we want!" the President of the Magazine Division and Publisher shouted.

In vain did I attempt to interject that this was a random distribution. One could not pinpoint the common factor between the six worst sellers. I saw I had gotten across nothing. It was a rout. The President of the Magazine Division and Publisher asked me how long it had taken me to do the analysis.

Shostakovich must have felt the same way when Stalin asked how long it would take him to rewrite his entry for the Soviet National Anthem. Shostakovich knew that to say five minutes was imprudent, so, to be safe, he answered five hours. Stalin awarded the prize to Alexandrov. "Well, to polish up the analysis did take a substantial amount of time, but if you mean just getting the initial Gaussian, about twenty minutes."

He was pleased. As long as no major R&D effort was involved, we would get sales figures from four other magazines and repeat the exercise. Then the President of the Magazine Division and Publisher snarled something about "rocket scientists doing circulation analysis," told me he wanted a private seminar, picked up a donut, and left.

In the aftermath I finally put to Moeling the question that had troubled me for months: had anyone on Madison Avenue ever made a histogram?

"Kelvin was the fortieth person to have the idea for the refrigerator," he replied.

VIII

The meeting broke up. I never heard anything more of the private audience I was to give to the President of the Magazine Division and Publisher, or of the sales figures from the other magazines. Some time later another editor attempted to persuade me that a Gaussian distribution "doesn't mean much"; he reminded me that if sales were perfectly correlated with a single, randomly varying X-factor—which he intended to find—a normal distribution would result. I never heard anything more about that either. The Circulation Manager emeritus still believes that the below-average sale in March 1989 can be explained, and he will continue to search for the cause.

Events had made clear that a physicist has no business in the real world, and I eventually left Madison Avenue. Shortly after I announced my resignation the President of the Magazine Division and Publisher vanished, though I cannot say whether the two events were connected. It is probably too much to hope that even under the new regime *Scientific American* will lower its draw. In December 1989, some months after I left, a predominantly red animal cover appeared. I claim credit for instilling the Editor with enough bravery to try it. However, that birds of paradise spread their plumage in December shows that *Scientific American* still has not learned to design a proper scientific experiment; I am willing to bet it sold better than average. I am also willing to bet that someone in the circulation department has seen the figures and has been yet again impressed by the selling power of red covers. Ah, well. As I said, a physicist has no business in the real world. We now return you to your regular programming.

CHAPTER 2

Instruments of the Future, Traditions of the Past

In the spring of 1841, a twenty-seven-year-old instrument maker named Adolphe Sax abandoned the Brussels workshop of his father and, with thirty francs in his pocket, set off for Paris, fame, and fortune. Soon he renewed the acquaintance of one Hector Berlioz, who, at thirty-eight, was as famous for his polemics in the *Journal des Debats* as for his *Symphonie fantastique,* which had premiered to scandal nine years earlier.

M. Adolphe Sax of Brussels, whose work we have examined, wrote Berlioz in the *Journal, has without any doubt made a powerful contribution to the revolution which is about to take place . . . At the same time, he is a calculator, an expert in acoustics and, when necessary, a smelter, a turner and embosser. . . . Composers will be indebted to M. Sax when his new instruments are in general use. If he perseveres, he will get the support of all friends of music.*

The revolution Berlioz expected Sax to lead was that of instrument making, and at the time, the composer had good reason to expect Paris to jump to Sax's command: even before coming to that city, Sax had developed the best clarinet in existence. But the revolutionaries were disappointed, at least until 1848. Only a year after his first effusion Berlioz found himself writing in despair:

Persecutions worthy of the Middle Ages are inflicted upon Sax . . . His workmen are enticed away, his designs stolen, he is accused of madness, and driven to litigation. A trifle more, and they would assassinate him. Such is the hatred that inventors always excite amongst those of their rivals who can invent nothing for themselves. . . .

In this instance Berlioz's words proved prophetic, and by 1845 he might have added:

Now these knaves and scoundrels, who are as fit to fashion instruments as the baboons and orangoutangs of Borneo, have chosen not to sit at the feet of this master of acoustical art as grateful apprentices, but to eliminate him altogether. Not long ago they turned their quaint *talents to the production of a diabolical contraption, designed, no less, to roast Sax alive in his very bed. Only a too-short fuse, leading to its premature ignition, saved their unwitting target from literal immolation. Even more recently, a trusted employee paid a midnight call on his master. As he stood on the threshold, awaiting a reply to his knock, an assassin suddenly appeared from the shadows and stabbed the poor man through the heart, evidently having mistaken his victim for Sax himself.*

Thus began the career of Adolphe Sax. It ended fifty years later in obscurity after a lifetime of patent disputes, courtroom litigation, and, finally, bankruptcy. In the meantime he not only invented the saxophone but improved the clarinet as well as dozens of bugles, cornets, and trumpets. These saxhorns, as they came to be known, were of such superior design compared with their predecessors that a few of them survive to the present day as the tenor and baritone horns used in brass bands.

But history is a battlefield strewn with the corpses of forgotten novels, operas, and scientific theories. Neither are musical instruments exempt from extinction, and most of Sax's inventions come down to us as fossils in museum display cases. Even the saxophone has never found a permanent home in the orchestra. Berlioz, in his enthusiasm, expected a revolution from Sax; instead he got a modest evolution. Why do some instruments survive and others die out? What makes one instrument better than another? What ensures the survival of the fittest? While we are posing questions: What is the shape of a violin? Why does an oboe sound different from a clarinet? Why are brass instruments made of brass?

The answers to these riddles are sometimes silly and sometimes profound. They involve a subject that has been infused with as much superstition as science, as much hearsay as hard fact, and patent disputes aplenty. It is the science-art of musical acoustics, and its modern practitioners, one hundred years after the death of Sax, are using their knowledge of physics, computer modeling, and high-tech materials to bring instrument making to the verge of a new revolution. *Revolution? My good fellow, let us not be too hasty about this.* Agreed, Sir, let another century pass and we may know whether current research is leading to revolution, evolution, or dead end.

Instrument design has, frankly, not progressed very far since the heyday of Sax. The first reason for this is the perpetual "funds for fashion" dilemma: not having the obvious potential to create high-powered lasers or quark bombs, acoustics tends to attract music-loving physicists and engineers who are willing to carry out their investigations as after-hours hobbies. The second component of the viscous drag on progress is that "a true conservative never tries anything for the first time," which could well be the motto of the average musician—who trusts scientists about as much as the Bourbons trusted Napoleon. And finally, scientist-designers themselves have frequently regressed the field. In their analytical enthusiasm they often forget that the ultimate test of an instrument lies not in the equations but in the hands of a performer. (You have only to think of Avery Fisher Hall to be reminded of scientifically wrought acoustical disasters.)

One scientist with a sympathetic ear for musicians is Carleen Maley Hutchins, who has led the recent struggle to turn the art of violin making into a science and who is known to many enthusiasts as the author of several *Scientific American* articles on the subject. Trained as a luthier as well as a biologist and a physicist, she has been testing and building violins for almost forty years in an effort to discover what distinguishes the good violin from the bad, and the great from the merely good. The work has been along a fairly lonely road; because she is not formally connected with a university, Hutchins's research is funded entirely from philanthropic sources and the sale of her own instruments. But recognition has come: the first silver medal for musical acoustics of the Acoustical Society of America, vilification in the press by paranoid violin makers—"conspirator"—and threats to burn down her house. *This superstitious rabble, which doesn't have the brains of a flea on a chain, should be buried in a granite mausoleum under the inscription "Cum mortuis in lingua mortua."*

Hutchins shrugs off the opposition, saying, "I must be doing something right," and her home has not been reduced to a pile of ashes. Indeed, when you walk into this turn-of-the-century house you find yourself surrounded by violins, violas, cellos, and basses that line the walls, chart recorder graphs heaped upon the dining room table, and a caged crow in the den that also serves as the office of the Catgut Acoustical Society, founded by Hutchins's mentor Frederick Saunders in 1963 to coordinate violin research.

Upon seeing her data on the table, my first question to her was, "Well, can you tell me what distinguishes a great violin from a merely good one?"

She replied flatly: "No."

Those of you reading this article to discover the secret of the Stradivarius can stop now. If only because wood varies so much from one piece to the next, it is impossible to make consistently great violins. "One thing they never tell you," says Hutchins, "is that Stradivari made some pretty poor sounding instruments." She will tell you (seriously) that subjecting a new violin to 1500 hours of WNCN will improve its quality. I suppose the next experiment is WABC.

Such levity aside, it is clear that Hutchins holds the greatest respect for violins and violin makers. "The violin is the most astounding acoustical phenomenon ever perpetrated by and on the human nervous system," she exults. "Whenever we have tried to deviate substantially from the proportions evolved by the great masters, we find the instruments get worse."

But then she adds, "Soon we will be able to surpass the Strad and do it consistently. I know that sounds brash, but I think it's true." If the statement is brash, its brashness is scientific. A scientist sees no reason to regard the evolution of the violin as having necessarily climaxed three hundred years ago with the work of the Cremonese giants Antonio Stradivari and Giuseppe Guarneri. For instance, several aspects of the violin seem somewhat arbitrary. Its body length, now standardized at 14", may have been chosen because the predominant seventeenth-century string material was catgut. For a given pitch, a longer string must be tightened more than a shorter string, and a violin length much greater than 14" could cause the string to snap. Furthermore, Stradivari appears to have based some of the proportions of his instruments on the famous "golden section." [Two numbers appear in a golden section if their ratio is $\frac{1}{2}(\sqrt{5} \pm 1)$.] While this may have resulted in a miracle for violin making, its basis can only be considered numerological.

Nevertheless, certain refinements allowed the violin to project better in large concert settings than older instruments (see box), and it eventually eclipsed its competitors. Actually, when the Stradivarius was introduced, its tone was considered loud and harsh compared with the softer and sweeter violin called the Amati (after Strad's teacher). Only one hundred years later, when larger concert halls came into use, did the Strad win out—and not without modifications. Most come down to us with lengthened necks, increasing the power output. Another good reason not to worship at the altar of the sacred Strad—it is already defiled.

Today, with the ever-increasing size of auditoriums, the same environmental pressure confronts violinists. Can anything be done about it? Hutchins has long been involved in designing new species of violins. The first step in this process is to understand in more detail the physics of already-existing violins.

To a physicist, any musical instrument is an oscillating system. Imagine a violin string fastened to a peg at each end. If we pluck the string it can vibrate in many resonance patterns or modes (Figure 2.2). But any allowed mode must be such that the displacement of the string disappears at the ends—obviously, since these are fixed. Places where the string displacement is zero are called nodes, and places where the displacement is a maximum are termed antinodes. Now, the lowest mode of oscillation, as it is called, occurs when the string displays one hump halfway between the two ends, or, more

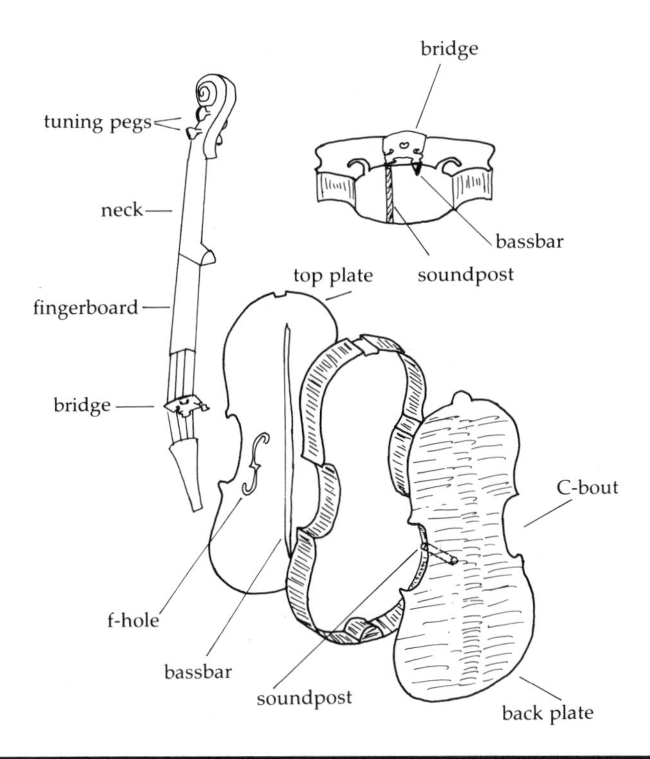

The early history of the violin is lost to the sands of time, but by the early sixteenth century an instrument emerged that closely resembled the one we know today. It also bore close kinship to the rebec, the Renaissance fiddle, the lyra da Braccio, of which no authentic examples survive and which the violin soon eclipsed. Why? Contemporary pictures of these instruments show them to have been made in various shapes—predominantly pear and oblong. By contrast, the violin is flat, generously rounded at the edges, and its most noticeable features are the indentations or C-bouts along the sides. The C-bouts evolved for practical reasons—they allow the bow easy access to the strings. As to why a violin is rounded (particularly on the inside) rather than boxlike, Carleen Hutchins explains that sharp corners trap air and reduce sound output.

The most important advance over its predecessors may have been two small pieces of wood that the rebec, at least, evidently lacked. When you look at a violin from the outside, it appears perfectly symmetrical. If you peer through the f-holes, however, you find you have been deceived. Fastened to the underside of the top plate and running parallel to the lowest string is a piece of wood called the bassbar. On the opposite side is no bar but a small wooden pillar, known as the soundpost, which is friction-fit between the front and back plates. Its position is very crucial. The soundpost does not, as many people believe, couple the top and back plates to enhance sound production. As the physicist Felix Savart discovered in the nineteenth century, you can clamp the plates on the *outside* above the same positions where the ends of the soundpost touch on the inside, and the effect does not change. No, together with the bassbar the soundpost makes the violin *asymmetrical*. The asymmetry is necessary to get maximum sound out of the instrument. If the violin were as symmetrical as it appears, then, under the rocking motion of the bow, first the air above one f-hole would be disturbed, then the other—180 degrees out of phase. So the oscillations would, to a good approximation, cancel, and less sound would escape the instrument. "Whoever originally moved the bassbar to one side of the instrument was a genius," says Hutchins.

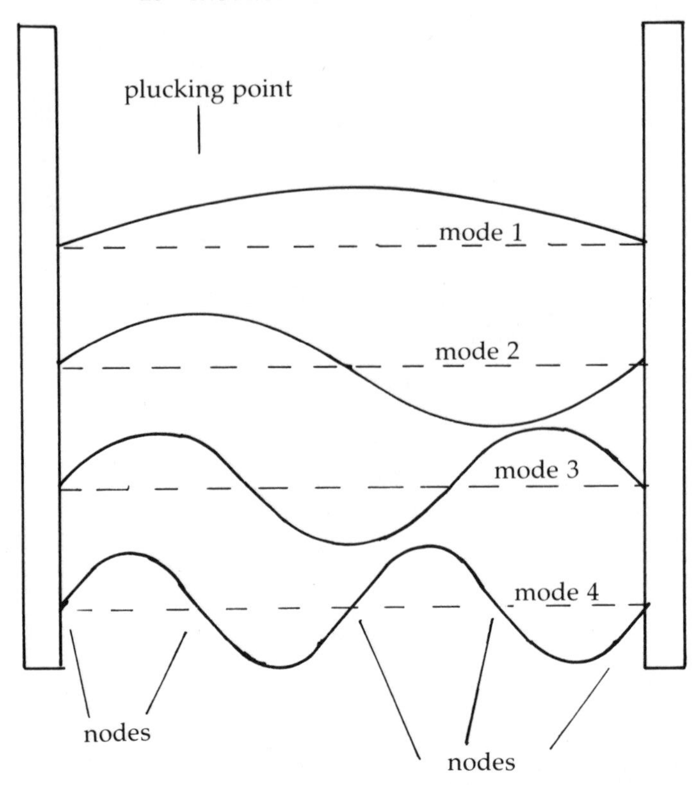

Figure 2.2 The four lowest modes of a violin string plucked at one-quarter the distance from the fastening point. The first mode, called the fundamental, shows one crest between the ends. The second mode, called the second harmonic or first overtone, displays two crests, and so on. Places where the string does not move (for instance, the ends) are termed nodes. Points of maximum displacement are termed antinodes.

technically, when the wavelength of oscillation is twice the length of the string. From a basic relationship between wavelength and frequency, the lowest mode gives the lowest frequency of the string, known in music as the fundamental. The second allowed mode occurs when the string displays two humps between the ends, and the frequency of this oscillation is twice that of the fundamental—a musician would say an octave higher. The third mode takes place at a frequency three times that of the fundamental, and so on. For violin strings, all allowed modes take place in integral multiples of the fundamental frequency. In musical terms, we say they are harmonically related, a fact that will turn out to be of the utmost importance. As an additional bit of terminology, for harmonically related modes, the frequency of the second mode is also called the second harmonic or first overtone.

Of course, a real violin is infinitely more complicated than a string stretched between two pegs. The vibration of the strings is transmitted through the bridge to the body to the air inside and thence to the audience. A real violin is so acoustically complex that to this day there does not exist a theory that entirely explains the behavior of the instrument. Research has therefore been highly empirical; the physicist, like the luthier, does an experiment, finds that such-and-such a modification improves the violin or makes it worse, but lacks a theoretical explanation. *It is impossible to adequately express our fervent hope that the next generation of scientists shall turn their formidable talents to the solution of this ever-vexing riddle.*

For her part, Hutchins considers that three important empirical developments regarding conventional violins have emerged from Catgut Acoustical Society research. The first of these has come to be called "mode tuning" and was suggested first in 1830 by the physicist Felix Savart, who asked: "What sounds ought the top and back of a violin have before they are joined?"

With the help of his friend Ernst Chladni (pronounced "Kladni"), Savart eventually gave a partial answer to this question: the sound of the top plate of a good violin varies between third octave C♯ and D, and for the back plate between D and D♯, so that the sounds are between a halftone and a whole tone apart. What did Savart mean by this?

Every freshman physics student has played with Chladni plates in the lab. These are thin metal plates, circular or square, that are typically clamped in a vice via a wooden handle bolted to the plate's center. When you bow the plate, it oscillates in much the same way as a violin string. But, unlike the string, the plate is two dimensional, and thus the allowed modes of oscillation are much more complicated. The student can observe the nodes and antinodes by sprinkling the plate with salt or powder; the moving parts (antinodes) bounce the salt into the stationary parts (nodes), where they remain, creating delightful designs known as Chladni patterns.

A number of scientists, including Hutchins, have followed Chladni's example and performed the same experiment with violin plates. Hutchins places a top or bottom plate on foam rubber supports at the plate's nodes; it is thus lying a few centimeters above a loudspeaker, which is itself wired to a tone generator. By adjusting the frequency of the signal, she can excite the various modes of the violin plate. Each mode generates a characteristic Chladni pattern, which she can visually identify by sprinkling the plate with Christmas tree glitter.

More accurate measurements, down to wavelengths of light, can be made using laser interferometry, but this technique is not necessary, being a "severe case of overkill," as Hutchins explains. (After all, Strad used only carpenters' chisels, and his senses of touch and hearing.) Over the years, Hutchins has decided that the unattached plates of the best violins exhibit the first mode at around 92 Hz, the second mode at about 185 Hz, and the fifth mode at 370, these three modes being the most critical for "tuning" violin plates.

Now we can understand Savart's remark. When a luthier makes a violin, he holds the plate at a certain spot and taps it with his knuckle. Though he almost certainly does not know what he is doing in terms of physics, he is holding the plate at the crossing point of the nodes for modes 2 and 5 and is therefore exciting the corresponding frequencies—which should ideally be an octave apart. If this taptone pitch is off, the luthier, by virtue of experience, scrapes a little wood from the proper places to tune the plates. Originally Hutchins confirmed Savart's findings that the taptones of the best violin plates should be a halftone apart between top and back. But recently she has decided they should be identical. She attributes the discrepancy in findings to the probability that Savart used violins with smaller bassbars than are used today.

I should say that doing the taptone test by ear requires some skill; Hutchins allowed me to test one of her plates in this way and all I heard was the sound of my knuckle on wood. The glitter test makes life easier for luthiers who lack absolute pitch, and Hutchins maintains that "even some violin makers who oppose scientific techniques in public are using the glitter test in private." Despite the cry of these two-faced reactionaries, scientific mode tuning appears to be gaining popularity, and Hutchins frequently travels to give seminars on the subject.

A second Catgut discovery that Hutchins deems important she refers to as "matching the beam mode and lowest cavity mode" of the violin. Over the last fifteen years, in addition to her other work, Hutchins has measured the response curves for several hundred instruments. A response curve is merely a graph of the amplitude or loudness of the violin output (in decibels) versus the frequency, essentially like the graphs that accompany audio cassettes (Figure 2.3). Unlike a good audio tape, any violin response curve will contain a forest of peaks and valleys that indicate where the violin responds well and where it responds poorly. Some of these peaks and valleys point to various mode positions.

The so-called beam mode for a violin lies between 260 and 300 Hz and represents a bending of the entire instrument across three nodal

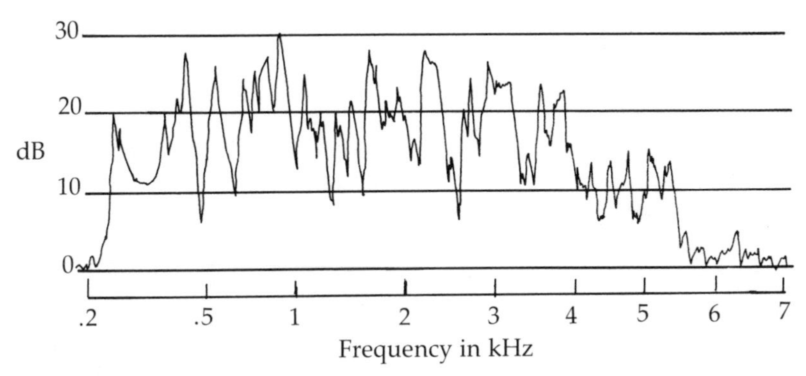

Figure 2.3 The response curve for a 1715 Stradivarius (after Saunders). The sound intensity, measured in decibels, is plotted against the frequency, measured in thousands of cycles per second (kilohertz). The jagged appearance of the graph shows that the violin resonates better and hence produces sound more easily at some frequencies than others. Certain of the peaks and valleys point to mode positions.

lines: the line that traverses the widest part of the body, a line just below that formed where the neck joins the body, and a third line near the tuning pegs. The lowest cavity, or Helmholtz, mode represents the motion of the air within the violin as it is first exhaled and then inhaled through the f-holes under the rocking motion of the bow. In fact you can put your face to the instrument and feel the 10 mph wind.

After analyzing two to three hundred instruments, including several Strads, Hutchins has concluded that in the best instruments the lowest cavity mode occurs at the same frequency as the beam mode. When I asked her if she had a theoretical explanation for this, she told me to get to work on it. That project remains untouched, but it is fairly plausible that when two major resonances coincide, the violin will spring to life.

Hutchins's third discovery also concerns the relationship between modes—in this case, the frequency difference between the first top plate mode (the "B_1" mode) and the second cavity mode (the "A_1" mode). If the frequency difference between the B_1 and A_1 modes is about 100 Hz, the violin will have a harsh sound. In violins used by concert soloists the difference is 50 to 60 Hz, and in chamber instruments the difference goes down to between 10 and 20 Hz. Hutchins considers this work her *magnum opus*, and it may be that in this discovery lies the distinction between the merely good and the great violin.

So far we have dealt with conventional violins, but the Catgut researchers have not stopped there. Hutchins and her colleagues have used scientific reasoning to produce an entirely new family of violins, which she considers the fourth major achievement of their labors. This violin octet, which consists of eight violins ranging in size from the treble to the contrabass, is not only the first new species of violin in centuries but shows directly how science can influence the design of musical instruments.

To understand the motivation for the new violin octet, we must first delve again into musical history. Before the nineteenth century, the violin, viola, cello, and bass competed actively with another string family, the viols (pronounced "vials"). The viols enjoyed several advantages over the violin. First, all were played in the same manner, allowing the musician to perform on the treble, tenor, or bass viol with equal facility, something impossible to do in the conventional string quartet. Second, the viols, being scaled versions of each other, all possess nearly identical tone qualities or timbre, which the conventional violin family lacks. (The muffled performance of the viola, and to a lesser extent the cello, is a frequent complaint of musicians and audiences alike. Scientifically this is due to the fact that their body size is smaller for their tone range relative to that of the violin.) Thus ensemble playing was rendered much more homogeneous by the viols. Nonetheless, the large concert hall problem of the nineteenth century edged the softer viol into extinction.

Ideally, one would like to see a scaled version of the violin; this would combine the homogeneity of the viol family with the projecting power of the violin. Thinking naively in terms of the oscillating string example given earlier, you might suspect that you could construct a violin an octave lower than the standard by doubling all the linear dimensions of the instrument. This is basically true, but if you wanted to build a contrabass violin in this way, you would quickly find that the length of the body should be six times that of the violin, or 84", and its width should also be six times that of the violin—equally ridiculous. That the body of the real bass is only about 43" indicates that it is not scaled in an obvious manner from the violin. Actually, it is not a member of the violin family at all, but an historical holdover from the early viols.

At least two attempts have been made in the past to scale up the violin. Michael Pretorius, in his *Syntagmatis Musici* of 1619, describes a family of violins very similar to the one Hutchins has created, though no examples of it survive and she was unaware of it until after her octet was created. A better example is a dinosaur known as the octobasse, apparently designed by Savart for Hector Berlioz; one of these is now housed in the Paris Conservatoire. The monster stands twelve feet tall and may have been the result of linear scaling. *My astonishment at hearing the vast, subterranean rumblings of this cyclops was extreme in the least. But, as you may well imagine, the instrument presents some severe practical difficulties. Two men are required to tune it, one on the second floor, through which a hole has been cut to accommodate the creature's neck, and to play it involves a dance on three pedals to manipulate the mechanical fingers, an activity uncomfortably reminiscent of musical hopscotch.*

The modern violin octet avoids such elephantiasis by proceeding from sound scientific principles. Many systems in physics obey the same sorts of equations; it turns out that masses on springs and pendulums behave very similarly to resistors, capacitors, and inductors in an electrical circuit. If you envision the vibrating plates of a violin to be composed of little masses on springs, then it is not too great a leap of faith to imagine that you can model a violin as an electrical circuit. This is exactly what John Schelleng did in his 1963 paper "The Violin as a Circuit." The resulting equations—known as ordinary differential equations—allow you to read off what the frequency of the violin-circuit should be in terms of its length, height, thickness, and mass. Thus you have derived a set of *scaling laws;* pick a frequency and the scaling laws tell you how big and heavy you must make the instrument.

When composer Henry Brant requested a new violin family, Hutchins collaborated with Schelleng to build the octet. It is based on Schelleng's scaling laws and succeeds in combining the best features of the viol and the violin. Her larger instruments are truly scaled versions of the violin and so should properly not be called cellos and basses, though two of them approximate the cello and bass in size. Like the viol, the new violin ensemble provides a homogeneity of sound that the standard string quartet does not match. Furthermore, since scientific scaling has been employed, the resonances have been kept at their proper frequencies for each tone range; this eliminates the muffled viola problem mentioned earlier. Even the violin has been redesigned to provide not only good sound quality but also greater tonal output, thus making it equal to the largest modern concert hall.

Some years ago, Hutchins reports, the cellist Gregor Piatigorsky exclaimed "Bravo!" upon playing the "cello," and she tells about the phone call she received the day after her "viola" debuted in New York. The caller, who spoke in broken English, was Leopold Stokowski, who said, "I want one of those for my orchestra!"

Not content with hearsay, *Discover* invited four members of the

Philadelphia Orchestra to appraise Hutchins's work: Nancy Bean, violin; Irving Segall, viola; Lloyd Smith, cello; Roger Scott, bass. Each tried several instruments in different rooms of the house and at first did not compare notes.

The phrase "very responsive" was used to describe every one of the instruments, as was the adjective "raw." Bean, who normally plays on an 1875 Lyon Silvestre, said of the first of Hutchins's conventional violins, "It's very easy to play and projects well. A little raw but that's probably because it's new. It could develop into a very fine instrument—all the right qualities are there." Incidentally, though she did not know it at the time, the two specimens that she preferred were those subjected to 1500 hours of WNCN.

Segall, who in the orchestra holds forth on a 1936 Italian Carletti, was impressed enough with Hutchins's conventional viola that he would consider testing one in the orchestra. "The mechanics are all there; the question is whether you like the tone quality of this instrument." Bean also mentioned that the Hutchins viola compared favorably with her 1939 Czechoslovakian Vsoky, on which she frequently performs.

As for the two cellos he tested, Smith pronounced them "competent instruments." He did not prefer either to his own 1686 Grancino and felt that they would not be suited for large halls, but Hutchins's more recent effort, in particular, "had a useful range of good color and would be quite suitable for chamber music."

The bass on hand was actually one of the octet's "contrabass violins," which is noticeably larger than the ordinary double bass and requires a readjustment of finger positions. Scott, who plays on an eighteenth-century Italian D'Aglio, frankly had difficulty negotiating the larger dimensions and finding the correct finger positions. He also noted a certain rawness and found "a surface sound on the top three strings, though the bottom string was very resonant."

By contrast, Smith was most satisfied with the octet's "cello," or baritone violin. He adjusted to it with apparent ease, found that it projected well, and called it "great fun to play." About the gamba-sized tenor violin he had similar remarks. "They're all so responsive." Segall, on the other hand, felt uncomfortable with the octet's large "viola" under his chin and would have to learn to fiddle the instrument on his knee—as in fact it was designed to be played. When Bean took up the tiny treble violin—strung with carbon rocket wire, the tension is so high—everyone laughed. But partly in surprise; its tone proved unexpectedly sweet.

With the musicians gathered in the living room, Hutchins used the prototype of her 15" violin to demonstrate the second Catgut discovery. She placed a piece of clay on the end of the fingerboard to align the cavity mode with the beam mode, which on that specimen had been deliberately left out of alignment. She handed it to Bean, who played on it for a few minutes; then Hutchins removed the clay to unalign the modes once more. Bean and everyone else present jumped at the difference—what had been a large, powerful sound was now weak and puny. Satisfied that scientists know a few things, the quartet got down to the serious business of playing Mozart—on one of the most peculiar assortments of instruments ever assembled.

It would be too much to claim that the members of the Philadelphia Orchestra were ready to trade in their priceless strings for Hutchins's experiments. A proper trial run would have lasted six months. In that regard, Smith summed up the afternoon best: "The survivability of these instruments depends on whether you can find someone to champion them, as Brahms championed the clarinet." For Smith's part, he would be glad to take the baritone on tour and hopes that Hutchins will approve.

It would be strange these days if computers did not get into the act of violin design, but rocket engineering comes as a surprise. I mentioned earlier that a scientific luthier uses the glitter test to tell whether a given vibrational mode is at the proper frequency. If not, the craftsman scrapes a millimeter or so from the right places to adjust the mode. To know what frequency you are aiming for is one thing; to know how and where to scrape to get there is quite another, and the eight-year apprenticeship of the luthier trains him intuitively to do exactly that. Another member of the Catgut Society, Oliver Rodgers, intends to compress eight years into one half hour. Rodgers is an adjunct professor of engineering at the University of Delaware, who first came to violin research when he took apart his daughter's violin, scraped, failed to improve the performance, and enrolled in Hutchins's mode-tuning course.

Now to study the vibrational behavior of a violin he has adapted a computer program originally written by the Systems Development Research Corporation to study the vibrations induced in satellite boosters by the uneven thrust of the rocket engines. Called SUPERB, the package uses "finite element analysis" to simulate a rocket or violin by dividing it into small blocks which move according to standard Newtonian mechanics. Rodgers first specifies the type of wood by inputting the nine parameters needed to characterize the elastic properties of any material. Then he can adjust the thickness of the plate to simulate the luthier's shaving and can watch the modes change position. In this way he can predict what changes any tinkering will have on the fiddle and thus guide the craftsman. His code, which considers only the back plate of a violin, takes 1 to 2 hours to run on a VAX 780 computer, but probably less time on newer machines. Similar work is apparently being conducted on the Cray-1 at the Naval Postgraduate Center in Monterey, which leads you to wonder about the role of violins in the nation's defense—can music soothe the ICBM? Although Rodgers's results are preliminary, Hutchins says they have already proved useful to her in fashioning instruments.

Rodgers would like to eventually model the entire violin. Now he is working on the top and the bridge. In regard to the latter, virtually all violin makers buy bridge blanks of specially treated maple—treated, some say, with rabbit urine (I was not joking when I said this field is laden with superstition). The luthier must then whittle down the blank to the intricate design seen on a finished violin. "Where to whittle is a black art," Rodgers explains; "Whether you whittle below or above the knee makes a large difference." Rodgers hopes to use SUPERB soon to direct bridge whittling, though it is not likely he will be able to model the effects of rabbit urine.

Closely related to the work of Rodgers is that of Kenneth Marshall at the BF Goodrich Research and Development Center in Ohio. Marshall comes to the violin via his background in vibrational and modal analysis, which during working hours is applied to automobiles and tires. Unlike Rodgers, Marshall has concerned himself with the vibrational modes of real violins. He supports a fiddle on a test stand by rubber bands attached to the C-bout corners. A tiny accelerometer is waxed to the point of interest, and then the instrument is tapped with a small impact hammer that contains another accelerometer. The procedure is repeated at 190 locations around the violin so that a complete picture of the vibrating instrument can be made. Marshall finds no less than 35 bending modes below 1300 Hz for the violin, which are now available for viewing by fans on a computer-generated video. From the purely scientific point of view, an important finding is that the bending modes are *linear*, as opposed to nonlinear. Loosely speaking, this means that the system vibrates, but not so much that it becomes impossible for the physicist to understand. From the practical point of view, Oliver Rodgers has used Marshall's video to make sure his simulations are simulating something that exists. And Marshall's video also confirmed to Hutchins that the first beam mode of a good violin should match the air mode—the second significant discovery of Catgut research.

An even less traditional approach is taken by Daniel Haines, who plays the violin and directs the graduate engineering program at

Manhattan College. Noting that the intrinsic variability of wood is a prime cause of the inconstant quality of violins, he would like to get rid of it altogether. The difficulty in finding a substitute is that spruce, from which violin top plates are invariably made, is a very unusual material. The theory of bending plates tells you that you need to duplicate the surface density of the spruce, the stiffness along the grain, and the stiffness across the grain (which in spruce occur in the extreme ratio of 10 or 20 to 1). Experience tells you that the rate at which vibrations damp out along the grain and across the grain is also extremely important for the tone of a violin. So to reasonably approximate spruce you need to duplicate at least five characteristics. You cannot do this with any other single material, but Haines found that a laminate of graphite-epoxy and resin-impregnated cardboard comes very close. In 1975 he engaged Hutchins to build a prototype, and the first violin with a graphite-epoxy top plate was born.

Soon after Haines announced his results, the Ovation Guitar Company began to manufacture their best-selling Adamus (Greek for diamond) guitar on the same principle. In response Haines lodged an interference suit against Ovation. Owing to slightly different details of the laminate, the law was able to resolve the dispute in a Solomonic fashion: recently a patent was awarded to Ovation for their laminate and one to Haines for his.

Haines hastens to point out that this is a student instrument, one designed to be consistent and affordable, and is not intended to compete with the Strad. Hutchins adds two more difficulties: "It's not very pretty, since graphite-epoxy is black, and *I'm* not going to build the things. The damned stuff festered under my skin for two years." There is a third difficulty: *Where is that Orpheus who gazes at the horizon instead of at his feet who would brave critics and public alike with a synthetic lyre?*

Nevertheless, we asked Nancy Bean to try it. The sight of the black top plate produced a grimace, but Bean's expression soon changed. "It's a violin!" she said, officially bewildered. Roger Scott, who was then listening, agreed that "behind a screen you'd never guess this wasn't an ordinary violin." And Irving Segall said that he "frankly found it unbelievable." Scott adds that the graphite hybrid would indeed be very useful in schools where "students give instruments a rather more rugged treatment than do professionals."

Score one for physics.

I have yet to mention the most controversial technique of all: varnish. In every article on violins the author is obliged to expound upon the mystical properties of Stradivari's varnish. Sea water, dragonfly wings, bumblebee skeletons, shrimp shells, God knows what else, has been claimed for the Secret of the Stradivarius. *Varnishers are like those unfortunate souls in possession of a piece of the True Cross who are condemned to wander eternally across the face of the earth accompanied only by the vacuous bliss of delusion.*

A few years ago a Texas biochemist, Joseph Nagyvary, captured the cover of *Science '84* for his varnishings, as well as prominent spotlights in the *New York Times* and other newspapers. I leave it as an exercise for the reader to ask responsible acousticians and violin makers for their opinion of Nagyvary. Could *Science '86* have folded because of this article? Every study of varnish indicates—and most violin makers will agree—that while varnish protects and lends a beautiful finish to the instrument, it diminishes the violin's ability to vibrate. In other words, he who varnishes best varnishes least.

Requiescat in pace.

The shadow of neglect that the more glamorous strings have traditionally cast upon woodwinds and brass extends to research on these instruments. No Split Reed or Swollen Lip Acoustical Society exists to advance our understanding of the descendants of Pan and Syrinx. And the acoustical ignorance that accompanies such neglect has frequently served to regress, not advance, their development. This is a point not widely appreciated, but often stressed by Arthur Benade, director of Case Western Reserve's musical acoustics lab, and the country's leading investigator into the physics of woodwinds.

He notes that during the industrial revolution, when precision machining became a reality, instrument makers began to take pride in a clean, sharp, machine-tooled appearance. But sharp edges cause turbulence in the airstream of a woodwind, and that means energy dissipation and decreased output. Thus round edges are vastly preferable. The point was dramatically demonstrated to me during a recent visit to Case Western, a visit hosted by Benade's postdoctoral researcher and trumpet player Peter Hoekje. He handed our guest flutist, Robert Dick, two plastic flutes, identical except for the way in which the tone holes were drilled. Dick played each in turn, and the difference can only be called astounding. The smoothed-out flute projected a large, clear tone, while the sharp-edged flute sounded weak and exhibited lots of extraneous hissing noises. WARNING: DO NOT RUIN YOUR $3000 INSTRUMENT BY TAKING A FILE TO IT.

Another instance in which science hindered, not helped, came about with the introduction of the electronic tuner known as the Stroboconn. Captivated by the new technology and the objectivity of science, some instrument makers began to pay exclusive attention to instrument tuning and ignored the performers' requirement of good tone quality. The resulting instruments could be played in tune with poor tone or with good tone but out of tune. A generation of unplayable woodwinds was the fruit of this blindly scientific approach.

As a consequence of relative neglect and techno-misadventures, many aspects of woodwind physics are still poorly understood, and questions that were debated 150 years ago still have no definitive answer. Nonetheless, under Benade's direction, the group at Case has won significant ground toward improving the woodwind. One advance may be called mode alignment and is reminiscent of Hutchins's work on violins.

Although the reed produces overtones in integral multiples of the fundamental, the natural frequencies of the air column itself need not necessarily have harmonic relationships, owing to complex details of the bore shape. Specifically, the second mode may not be at twice the frequency of the first mode. The closer you can get to the ideal factor of two (for most woodwinds), the easier and better the woodwind plays.

In the clarinet or oboe, nonalignment of modes 1 and 2 causes problems because the reed then does not know at what frequency to vibrate to sustain the oscillations. As an analogy, think of pushing two people on swings, one swinging twice as fast as the other. If you push at the frequency of the slower swinger, you will keep both people going, though you will catch the faster swinger only on every other swing. But if their swinging frequencies are not quite two-to-one, it becomes difficult to find the proper rate at which to push. If you are not careful, you will get out of phase and either find yourself flat on your back or find that everything quickly grinds to a halt. In the woodwind, this problem manifests itself, for instance, as intonation drift during crescendos and decrescendos, something noted by Berlioz in his treatise on orchestration.

Benade has developed simple tests the player can perform to determine whether a given instrument's modes are properly aligned. Joining us at Case was Bob Hill, a faculty member at the Cleveland Institute of Music and solo clarinetist of the Cleveland Chamber Players. Hill, a progressive, joined us happily, saying, "This business is so stuffy that people use reed rush because they've never heard of sandpaper." Using Benade's test on two of Hill's clarinets, Hoekje confirmed good mode alignment on Hill's preferred instrument and poor alignment on the second. For the musician considering the purchase of an instrument, this test can considerably shorten the evaluation time needed to choose a good instrument.

Such tests can also guide the instrument maker in producing an instrument free of alignment error. To do this may require adjusting the size of the tone holes, and changing the shape of the bore and the bell. The important point is that since woodwinds are acoustically simpler than strings, all these changes can be calculated from fundamental principles. Benade has thus been able to design from scratch a new clarinet that has more stable intonation and is easier to play than the conventional model. Unfortunately, it requires a slightly different technique from the usual, and for that reason most clarinetists will not touch it.

Brass instruments also exhibit mode alignment problems, though to adjust them you do not redrill tone holes but rather adjust the shape of the tube itself. (For other intonational problems in brass, you can change the valve positions as well—a technique that resulted in some of Sax's more spectacular disasters.) William Cardwell, who was with Chevron Research in California for forty-one years, is the inventor of the trumpet analogous to Benade's clarinet. He explains that from the 1920s onward, Henri Bouasse and successors attempted to calculate the shape a trumpet must be so that its modes are in harmonic ratios. They all considered just the bell and failed. Cardwell took into account the mouthpiece and leader pipe and showed that if a mathematical curve known as a catenoid is used, you can produce an essentially ideal trumpet. He patented his invention in 1970, and the Olds Company went into production. But during the rash of 1970s mergers, Olds was bought out by a multinational, and after a few hundred Cardwell trumpets were built, the project was killed. Now the renowned trumpet maker Cliff Blackburn is manufacturing a few Cardwell-designed instruments.

Despite such sad tales, the work of Benade and his colleagues has influenced at least one scientifically inclined musician, Robert Dick. Dick is America's leading avant-garde flutist and is known for his advocacy of "extended techniques." Multiphonics, whispertones, circular breathing . . . all are featured on his latest album, *The Other Flute,* and explained to the interested student in manuals available from the Multiple Breath Music Company. Of Benade's work, Dick says, "I've come to regard the entire flute-body system as a resonator, and I think carefully about how sounds are produced."

He also experiments relentlessly. Dick is currently engaged in a collaboration with Albert Cooper, the world's foremost flute designer, to totally revamp the flute. This is a rather bold idea, because in its present form the flute is technically the most successful of woodwinds. The one unarguable nineteenth-century revolution in woodwind making was due to Theobald Boehm, whose 1847 flute so eclipsed its predecessors in performance of new music that it remains essentially unaltered to the present day. But music has continued to change since 1847, and virtuosi like Dick are trying to break the bonds of tradition—a tradition that limits the performer to playing one note at a time. As far back as 1810, it was reported (by flutist George Bayr) that one can get several notes out of a flute simultaneously; the result was so shocking to Bayer's contemporaries that a commission was set up to investigate it! Nowadays, multiphonics, as such multiple tones are called, are more commonplace, though perhaps not much more appreciated. To the average listener multiphonics sound like horrible squawks, but in Robert Dick's hands they become haunting and evocative. Thanks to performers like him, they have become an integral part of the modern woodwind player's technique. Multiphonic fingerings, however, are difficult and finicky and sometimes do not even exist on conventional Boehm flutes. Dick's prototype is designed to overcome this dilemma. With tone holes spaced along the side, back, and top of the flute, he can finger multiphonics as easily as single notes. But "specialty instruments are never successful in the long run," says Dick. "I want this flute to be able to compete with the Boehm flute in traditional music—and the Boehm flute is hard to beat." At present, Dick considers his prototype to be "80 percent successful, mechanically speaking, and the last 20 percent is going to be tough."

At Case, Peter Hoekje demonstrated a technique that could be useful to players and possibly designers of multiphonic instruments. As with violins, it is possible to plot the response curve of a woodwind. Hoekje does this by connecting a simple pulse driver to the mouthpiece of a clarinet. He then generates a response curve on a spectrum analyzer that shows the frequencies at which the clarinet is resonating and those at which it is not. By trying various fingerings he can sometimes produce a pattern on the screen with a double hump in nonintegral frequency ratios. This shows that the instrument is resonating at two non-harmonically-related frequencies simultaneously—the sure sign of a multiphonic. In our presence Hoekje found such a fingering, disconnected the clarinet from the measurement stand, and tried it with a reed. Voilà, the world's first predicted multiphonic.

Such a diagnostic test could be of immediate benefit to oboists and clarinetists. On the flute, as Dick explains, multiphonic fingers are transferable from one brand to another. On clarinets and oboes this is not true. A composer will frequently specify a multiphonic fingering to be used on the Marigot oboe, and the Lorée player finds that it will not produce the required sound—physically, the given fingering does not admit the required resonances. Using the response-curve test, Hoekje says, it would be a simple matter to compile a multiphonic fingering chart that would allow one to translate from Marigot multiphonics to Lorée. It might also facilitate designing instruments analogous to Dick's flute. Both projects are worth pursuing.

Another area of woodwind research that has received some—but not enough—attention is the reed. This tiny piece of vegetable matter is the bane of all clarinetists, oboists, and bassoonists. A clarinetist will typically buy a box of reeds, throw most of them out, and rework the remainder until he comes up with one or two good ones. An oboist's fate is yet worse, and the professional may spend fifteen to twenty hours a week hunched over his workbench. Usually he starts from the raw bamboo-like cane, splits it, gouges it, shapes it, wraps it with thread onto the tube that is inserted into the instrument, and spends an hour whittling the reed to perfection. Depending on experience, perhaps three out of five are playable, and one may be excellent. A topflight professional will use a reed only once, then throw it away; others put up with a reed for a few days to a week. To buy them is prohibitively expensive—up to $10 a throw—and in any case, no oboist in his right mind would sell a good reed. Bassoonists are faced with similar problems, though their reeds, being thicker, last longer.

There have been a few attempts to transform this medieval tradition into something compatible with the twentieth century, but they have not been very successful. *Arundo donax* possesses mechanical properties which are evidently not duplicated anywhere else in nature. The situation is desperate enough that when he heard we were in possession of a few synthetic reeds, clarinetist Hill cried, "I'd pay $10,000 for a reed that would last me the rest of my life."

Our first specimen was a clear acrylic reed made by the Bari Company in Fort Lauderdale. Hill pronounced it totally unacceptable, saying, "I wouldn't last a day in an orchestra with this." Nonetheless, a little sanding produced a noticeable improvement in tone quality. Hill was still dissatisfied and said he would never use it.

Our second specimen was donated by John Backus, professor emeritus of physics at the University of Southern California, who is author of the standard text *The Acoustical Foundations of Music* and a leading reed investigator. He explains that the natural vibrating frequency of a clarinet reed is roughly 2,500 Hz. If the reed vibrates at a much lower frequency it cannot adequately produce the high register. In an attempt to develop a synthetic reed with the properties of cane, he experimented with fiberglass and magnesium, but their resonance frequencies were too low. After reading about Haines's graphite-epoxy violin, he decided to try the same trick on clarinet reeds in order to mimic cane's density, cross-grain and along-the-grain stiffness. The result was a laminate of graphite-epoxy and microscopic plastic spheres.

Although Hill felt that the particular samples in hand were too soft for his tastes, the audience immediately applauded the vast improvement in tone quality and flexibility over the acrylic version. Hill exclaimed, "Why aren't these being developed?" Indeed, Backus reports that a New York firm once planned to manufacture his reed, but the venture collapsed.

If a clarinetist would pay $10,000 for a lifetime reed, it goes without saying that an oboist would pay $50,000. I can speak from personal experience, as it happens I am an oboist. For many years now the Fibercane Company has manufactured fiberglass oboe reeds. Although I have never heard of anyone using them, I procured a few of these for tests and was pleasantly surprised to find that I could produce a sound. The low register is almost acceptable, the middle range displays ducklike qualities, and the high register is impossible to support—probably because the natural vibration frequency of fiberglass is too low. Standard whittling tricks failed to improve the performance.

A version of Backus's reed for oboe does not exist, and so to date the lifetime reed remains the stuff of dreams.

We now leave the territory of hard science for slightly more speculative ground. I said earlier that many aspects of woodwind physics are not understood any better now than they were 150 years ago. To the scientist this is an annoyance; to the artist it allows for unbridled experimentation.

One perpetual riddle is whether the construction material makes any difference to the tone quality of wind instruments. Most musicians would unhesitatingly say yes. When the metal flute was introduced in the early nineteenth century, critics described its tone as "shrill," "harsh," "metallic." And how many times have you heard brass instruments described as "brassy"? Such adjectives are clearly astrological in origin; they make little more sense than to claim someone born under the sign of Scorpio has the character of a scorpion. Unlike the string instruments, where everything vibrates, in the woodwind the only thing vibrating—at least to a high approximation—is the air column itself. So to a physicist it seems unlikely that the material from which a woodwind is fashioned can make more than the slightest difference.

But the ear is more than slightly sensitive, and I confess to having been infected by the musician's conservatism. Some years ago I was in a position to buy a Lorée—choice of most American orchestral oboists. But this Loree was unusual in that the top third of the instrument was made of plastic. I balked, afraid that the tone quality would be inferior to that of the traditional grenadilla wood model. But side-by-side comparisons with several wooden specimens convinced me that there was absolutely no difference in sound between the two, and I bought the plastic Loree.

I recently asked Robert de Gourdon, the current proprietor of the Loree Company in Paris, about how the topjoint came into existence. Not surprisingly, the main motivation was to develop an instrument that would not crack, probably the chief worry of oboists and clarinetists. He says, "I walked into a chemical firm that is known for their work on plastics and handed a chemist a piece of grenadilla wood. I asked him if he could make me a material that duplicated its properties, and a few months later he handed me a piece of plastic." More than that de Gourdon hesitates to reveal, it being a trade secret.

De Gourdon agrees that there is no difference in sound quality but says that 99 percent of all oboists refuse to touch it. "If tomorrow you handed a violinist a plastic violin, do you think anyone would play it?" The plastic topjoint is not perfect, however. Owing to its surface characteristics, the water from the player's mouth adheres to the tube and clogs up the keys, sometimes making it impossible to play. The problem, as it turns out, can be solved by spraying the octave holes with ordinary silicone, making the plastic version in all respects superior to its wooden counterpart.

But why then have certain materials, such as grenadilla wood or ebony, become traditional in woodwinds? The noted American oboe maker Paul Covey in Baltimore—and virtually every other woodwind manufacturer—echoes de Gourdon in saying that machinability is the prime consideration when choosing a material. Indeed, woodwinds have been successfully manufactured from boxwood, rosewood, pear, cherry, maple, violet, cocus, metal, ebonite, and ivory. Sax did not build his saxophones and saxhorns from brass to make them brassy, but because he knew that wood was unsuitable for fashioning such long and convoluted tubes. *Yes, good fellow, but in your haste you forget that our shining musical firmament is still ruled by the stars, not science, and all but the most revolutionary artists will march to the grave with their rusted and cracked weapons.*

In the issue of material and timbre the flute may be an exception. The walls of the flute are thin enough—typically .016″—to vibrate, so it is conceivable that the construction material might influence the tone quality. In past epochs flutes have been fashioned from as many materials as oboes, or more; any distinguished flute collection even boasts a glass flute. Nowadays the standard material is of course silver, though some flutists prefer gold and even platinum.

In Robert Dick's opinion the best flutes today are to the 99th percentile indistinguishable. "But art takes place in the remaining 1 percent, so it is important to split hairs. I think the gold flute is overrated. There's a lot of social pressure here. If you're playing first chair in an orchestra on a silver flute and your assistant is playing on a gold flute that glimmers before the whole public, you've got to move up." He recounts that the world's first platinum flute, made by the Haynes Company for George Barrère, was built simply because Barrère wanted to own the most expensive flute in the world. As for the antique glass flutes: "Dead, horrible, lifeless."

Dick himself plays on the world's only stainless steel flute. He explains that the stereo, like the large concert hall of two centuries ago, has altered our listening habits. "With today's equipment (especially CDs), the listener is so conditioned to an up-close, present sound that when he hears a flutist in a real concert hall it sounds faraway and artificial. Unless the flutist can provide a real presence, he isn't going to make it." Dick says that the stainless steel flute, also designed by Albert Cooper, can be made thinner than a silver flute. This allows the flute to vibrate more and enhances the harmonic content, giving the flute greater projecting power. The next step, according to Dick, is to go to titanium, which would require the same advanced machining techniques as those wonderful titanium corkscrews sold on the Moscow black market—machining undoubtedly carried out after-hours at a strategic arms factory.

Dick is surely correct that fine flutes are to the 99th percentile indistinguishable. Also joining us at Case was Bickford Brannen, president of Brannen Brothers Flutes, currently this country's most successful flute manufacturer. Brannen brought with him a silver flute, a 9 carat gold flute, a 14 carat gold flute, and a platinum flute (prices upon request). Dick played each one in turn, including his stainless steel flute, while I sat with my back turned and made some notes. Next he played them in a different order and I tried to match the tones. I identified only the platinum flute correctly, which may have been chance. A poor ear for flutes? Perhaps. But Bickford Brannen tried the same test and also only came up with one correct match. You should not take these results too seriously; we were pressed for time, Robert Dick had not adapted to each instrument, and, in his words, "we had six winners." Lacking was a standard of badness that might have revealed a highly audible difference. And what is generally as significant to the player as the sound is the feel; not surprisingly Dick preferred stainless steel.

Juan Novo, a Miami flute maker, would probably agree with the test results. He claims that 95 percent of the tone quality is determined by the mouthpiece, and the material used to construct the body of the tube makes no difference. "You could make it out of paper." He has not gone that far, but he has recently created the world's first fiber-optic flute. Novo recounts that the idea for his "Fantasia," as he has christened it, came to him one night in a dream. The machining process was the subject of an NBC documentary, required three months, and "is a nightmare. You have to lathe it with cooling water so the acrylic won't deform, and keep a steady stream of pressurized air flowing to blow out the chips." The nightmare results in a rather steep price on the prototype—$15,000—but to jazz flutist Dave Valentin it is truly a fantasia, and he features the instrument on his ninth album, "Light Struck." That may be the real point: you can buy an LED light system to go with your Fantasia, and Valentin hopes to soon connect a color organ to provide a visual palette for his playing. As for its sound, Valentin says that the tone is a little softer than a metal flute. Regardless of its potential commercial success, however, the Fantasia's main role may lie in research. Because it is transparent, it should be possible to observe the airflow in the tube, and Novo says that plans are already under way to do such experiments at the University of Miami.

A similar development has taken place in the brass world. Ellis Wean, winner of the First International Tuba Competition and now tuba player with the Montreal Symphony, has recently patented a series of transparent mouthpieces for all brass instruments. Acoustically, Wean claims, the Tru-Vu mouthpiece makes no difference to the sound. It does, however, provide the teacher or student with an undistorted view of the lips. Once the player sees what his lips are doing, Wean claims, the learning process is vastly accelerated. As an added bonus, the lips of marching-band members won't freeze to them in winter.

Despite the previous warning that we have left the realm of hard science, you might feel that by the late twentieth century physics should have stamped the materials debate with its seal of objectivity. The matter is not so simple; the problem has too many variables. While an artist like Robert Dick knows from long experience that one flute differs in tone quality from another, it is much more difficult to find the cause of that difference, especially in materials. Brannen's gold flutes are not the same thickness as his silver flutes or his platinum flutes, and the total masses of the instruments differ. So it becomes impossible to isolate the variable that could cause a difference in tone quality. If you could construct a synthetic flute of the same density, thickness, and stiffness as a silver flute, you might well find they sounded identical.

By now every physicist reading this article will be exclaiming, "Why don't you measure the spectra of the various flutes already?" A spectrum of a note is simply a graph of the strength of each harmonic versus the frequency of that harmonic (Figure 2.4). It is commonly thought that the spectrum of each instrument is unique and that this is what distinguishes a given instrument from all others. Wrong. At least mostly. It turns out that the spectra of a note on the oboe sampled at two different points in an auditorium differ more than the spectra of the same note played on an oboe and a trumpet. The rise time of the note (the attack) and the decay rate may be more important for determining an instrument's tone than its spectrum. Science, at present, has not answered the question, and the ear is still

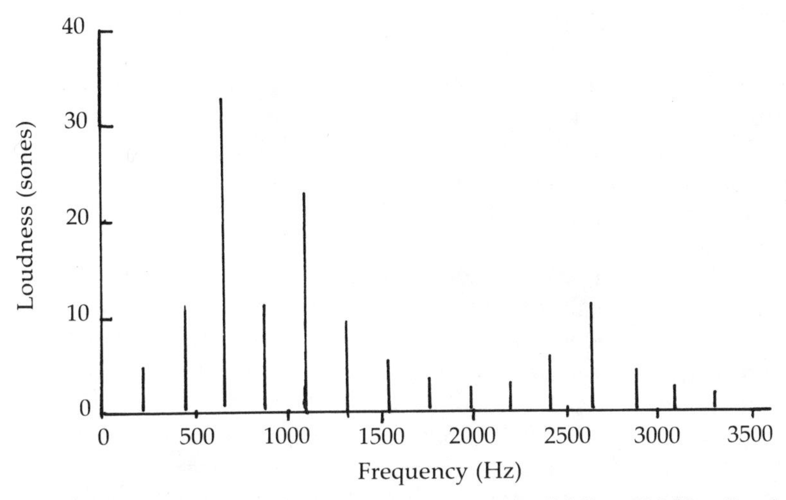

Figure 2.4 The spectrum of a human voice singing "ah" at 220 Hz. On the horizontal axis is shown the frequency of the fundamental and each overtone; on the vertical axis is shown the strength of the overtones in units called sones. All the overtones are at nearly integral multiples of 220. One sees that the fundamental is not as strong as the third, fifth, or twelfth harmonics. In principle, every instrument has a unique spectrum; in practice the spectra of radically different instruments can be almost indistinguishable.

the best judge. We were prepared to take spectra of the Brannen flutes at Case, but any results would have been meaningless.

For the oboe and clarinet families, if material does not enter into the tone quality debate, something else does—the reed. Because the clarinet uses a single piece of cane as a reed, while the oboe and bassoon use two pieces vibrating in opposition like a soda straw, most people assume that it is the single versus double reed that distinguishes the clarinet's wail from the oboe's plaint. This is mostly false. The primary source of the clarinet's tone is the cylindrical bore, which strengthens harmonics in only odd multiples of the fundamental. By contrast, the oboe's conical bore admits both even and odd harmonics. In fact, a small clarinet-like mouthpiece exists for the oboe when jazz saxophonists need to double on the instrument and fool the audience. They succeed.

So it is conceivable that the double reed might be replaced by a single reed, and perhaps both someday eliminated altogether. In the meantime, science cannot give a definitive answer to the most basic question of all—why one instrument sounds different from another.

One hundred fifty years in the past, Hector Berlioz described his ideal orchestra of 465 instruments. It included four octobasses, four tenoroons, three ophicleides, and five saxophones. I will end this essay with some speculations about the orchestra of one hundred years in the future. Will Hutchins's contrabass violins literally bring down the house? Or will they be displayed in glass cases next to Sax's 13-belled trombone? Will flutists play on multiphonic instruments and will clarinetists use Benade bores and Backus reeds? Can we imagine woodwinds that sound like synthesizers or strings that sound like tinkling bells?

At first you might be tempted to answer yes to most of these questions. Why should not Hutchins's violin be adapted? After all, it is only one inch longer than its predecessor; Dick's stainless steel flute is a few thousandths of an inch thinner than Brannen's silver flute. From the peanut gallery in Carnegie Hall you will not see a difference; you may not even hear one. And innovations have been made: ten years ago Albert Cooper introduced a new tone-hole placement to improve the flute's tuning. The difference between the Cooper flute and its predecessors is perhaps a centimeter over the entire length of the tone-hole network. Yet companies that have not adapted it—such as the once famous Haynes—are facing severe financial difficulties.

The moral is not that all revolutions are successful but that microscopic changes eventually produce macroscopic consequences. And there are at least three forces aligned against macro-revolutionaries: sociology, biology, and physics. The social obstacles to innovation in a conservative field should by now be clear. *Are they not? Perhaps, if you permit me, a visit to a distinguished ophthalmologist would be in order.*

But the forces of biology and physics are also large. You might, for instance, imagine adapting scales other than those we now use. The 31-tone scale, for instance, has some musical advantages over the familiar 12-tone scale—principally, the seventh harmonic appears as a note in the scale. But while aliens with a highly developed auditory system might prefer 31-tone music, most human listeners just find it out of tune.

That point is highly debatable, but this one is less so: our ears like harmonic sounds; they do not like the sound of trash cans overturned in back alleys. We seem to require overtones to be in integral multiples of the fundamental. Most percussion and all trash cans fail to meet this demand. Some years ago Benade proved that if you want to produce a harmonic scale from a wind instrument, its bore must be cylindrical like that of the flute and clarinet, conical like the oboes and saxophone, or "bessel" like the brass instruments. No others allowed. So it may be that all the ecological woodwind niches are filled. The fact that the saxophone was the last successfully invented instrument, 150 years ago, must tell us something.

Yet another clue to constraints on the evolution of instruments has recently emerged from the corridors of Case Western Reserve. The harmonics in the spectrum of a typical instrument are roughly equal in amplitude until a "critical frequency" and then begin to get smaller. If you plot the spectra of all modern instruments beyond the critical frequency, you find that they all fall off at about 18 decibels per octave. Instruments such as the sackbutt and crumhorn, which do not obey this law, have over the centuries joined the dodo. Benade and others speculate that the 18-decibel-per-octave falloff may optimize a certain kind of perceived loudness (*die Schärfe*) the ear requires. If instrument makers created a new instrument that did not obey this rule, our aural system might eventually reject it.

So it may be that the evolution of conventional instruments is nearing its termination, and the next epoch will indeed be the age of the synthesizer. I hope at least they develop one that a cellist might enjoy playing. But we haven't quite gotten there yet, and since I opened with the words of a musician, I will close with those of a scientist, Felix Savart, writing in 1819: *We have arrived at a time when the efforts of scientists and those of artists are going to bring perfection to an art which for so long has been limited to blind routine.*

CHAPTER 3

The Seven Arrows of Time

LOVE is an elusive concept. If you have ever attempted to describe it, you have probably fumbled for words like "bliss," "tenderness," "exhilaration," "agony," and "despair." And you remain dissatisfied because love is all of these and none of these. Time has much in common with love—it is a concept we live with daily and hardly understand. You can't see time, you can't hear it or taste it, and while you may have a lot on your hands, it can slip away in the blink of an eye. We all know that time flies unless it stands still. We can measure it, yet even when we turn back the clock we are nevertheless certain that time marches on.

Time marches on. If there is one thing about time that most people would agree on, it is that time moves forward. We see people born and become old. Only in the pages of the *National Enquirer* do 142-year-old men grow younger. A watch falls from your hand and shatters. We never see a mass of gears and ratchets spontaneously assemble itself into a watch and leap into our fingers—unless it is in a movie run backward for comical effect, and then we immediately say to ourselves, "This can't happen." One of the most fundamental aspects of human existence is that we distinguish between past and future, and know which way time moves. No one who has seen two snapshots of a man as baby and adult, or of the watch in hand and shattered on the floor, would hesitate to say which was taken first. And this is in fact how we tell time: by processes that proceed inexorably, irreversibly in one direction—*forward*. (I admit that disagreement might come from wearers of digital watches, who are perennially doomed to see only one instant at a time and lose track of forward motion. Could digital watches be responsible for the "me" generation?)

All this seems quite obvious and trivial until you realize that physicists have yet to agree on a theory that explains why time goes forward. In better terms: why is there an arrow of time?

The difficulty started in 1687, when Isaac Newton published his monumental *Principia* and unleashed on the world the science of mechanics. Newton's mechanics allows you to calculate the orbits of the planets, the trajectories of missiles, the behavior of billiard balls when they collide. In fact, Newton's mechanics basically idealizes everything as billiard balls: planets are large billiard balls; water and air are composed of microscopic billiard balls.

One of the strangest things about Newton's theory is that there is no arrow of time built into it. Suppose I project a film of the earth revolving clockwise around the sun. If I run the film backward, the earth appears to revolve counterclockwise. You might be tempted to say that this means time is also running backward, but you would be jumping to conclusions. If the film were originally *shot* backward, forward motion is counterclockwise and backward motion is clockwise. There is simply no way you can tell in mechanics which way time is running, which is future and which is past. In the parlance of physicists, *mechanics is time-reversible.*

When a physicist says that mechanics is time-reversible, he is tacitly assuming that it contains "zero defects" or, more technically, is "deterministic." If I know the positions of all the planets at the present instant of time and know their velocities as well, the equations let me predict *exactly* where they will be an hour from now or a million years from now. Then I can reverse the clock and all the planets will return *exactly* to their present positions. I can continue to let the clock run backward and know *exactly* where the planets were an hour ago or a million years ago. Once again I can reverse the direction of time—now forward—and the planets will return once more to their original positions. In Newtonian mechanics, the future is *exactly* determined, as is the past. It has zero defects; it is time-reversible.

The philosophers of the Age of Reason were very happy with Newton's mechanics because it turned the universe into a giant clock. God was removed from the picture; He was necessary only to start the clock off and then He could relax—every subsequent event was determined *exactly*. No free will whatsoever. All our actions have been preordained since the instant God pressed the stopwatch's trigger. Even Bobby Ewing's resurrection on "Dallas" was made inevitable, not by declining Nielsen ratings, but by events that took place 15 billion years ago. However, the situation is even worse. Since you cannot distinguish past and future in Newtonian mechanics, the idea of "cause and effect" disappears. To the man in the street causes always come before effects, but because I can run the cosmic movie backward, I can always transform causes into effects and vice versa. I can calculate the present position of the earth either by knowing the events that took place an hour ago or will take place an hour from now.

Since Newton, many theories have been based on mechanics: Maxwell's theory of electromagnetism, Einstein's theory of relativity, and quantum mechanics. All these theories have one thing in common: *None contain an arrow of time. All are time-reversible.*

How do we reconcile the fact that no arrow of time exists in our physical theories with the fact that we experience such an arrow in almost everything we do? This is one of the deepest questions in physics, and if you raise it among physicists, you will see ordinarily nondescript people rise through the spectrum of philosophy, metaphysics, anger, and embarrassment. As one Nobel laureate put it to me when I pressed him for an answer: "It's one of those very profound questions I try to avoid by not thinking about it. But that's not for publication." Einstein himself retreated before the riddle of time's arrow. In a famous letter penned upon the death of his oldest friend, Michele Besso, Einstein wrote to Besso's sister: "Michele has left this strange world just before me. This is of no importance. For us convinced physicists the distinction between past, present, and future is an illusion, although a persistent one."

Roger Penrose of Oxford University is one of the world's leading cosmologists. Originally trained as a mathematician, he has made major contributions to the theory of black holes, and for the last ten years or so has led his Oxford group in the development of twistor theory (twistors are mathematical constructions meant to be the fundamental building blocks of spacetime). Penrose is also the accidental father of "white holes," which he presented as a joke at a 1970 conference by projecting a diagram of black holes upside down.

Penrose has thought a great deal about the arrow of time. In fact, he has compiled a list of not one but seven apparently independent arrows of time in nature, all of which point forward and all of which are unexplained by current physics.[1]

The first of these is what Penrose calls "psychological time," and I have already alluded to it. Psychological time is our feeling that events move inexorably *forward*. In Penrose's words, "potentialities become actualities." The sensation is a universal one, but the White Queen at least would object: "It's a poor memory that only works backwards," she said to Alice. And she is correct; we are made of atoms and molecules that obey Newtonian mechanics and Einstein-

[1] Although Penrose's list contains seven arrows, it seems to me that only six can actually be independent, since the direction of one is arbitrary. I mentioned this to Penrose, and he seemed to agree with me. In addition, Stephen Hawking has persuasively argued that the psychological arrow of time (arrow 1) is linked to the entropic arrow of time (arrow 7). In order to recover a memory, a certain amount of entropy is generated in the brain. Therefore the increase in entropy gives the sensation that time is moving forward.

ian relativity, and so there is no reason we should not remember the future. Perhaps clairvoyants would claim to have that ability. For the rest of us, when memory works at all, it is sadly limited to events that have already taken place. The first arrow of time.

Arrow number two is slightly esoteric and refers to the so-called "retardation" of electromagnetic radiation. As mentioned earlier, Maxwell's theory of electromagnetism contains no arrow of time. At the beginning of the old RKO Radio films, the RKO logo depicted a radio antenna emitting radio waves into space. If you ran the film backward, the waves would appear to be converging from all directions onto the antenna. According to Maxwell, such behavior is perfectly acceptable, but it would look no less strange to us than the broken watch suddenly assembling itself, and indeed no one has ever observed the "RKO reversed logo." Nature seems to distinguish the outgoing case (known in physics, for reasons which need not concern us, as "retarded") from the ingoing case ("advanced"). The second arrow of time.

For arrow three we continue to climb up the ladder of esoterica, to quantum mechanics. Quantum mechanics is the science that describes the interactions of subatomic particles, but it does so in a fashion so contrary to intuition that physicists have been arguing over what the theory means ever since it came into existence sixty-odd years ago. If you calculated the outcome of a quantum-mechanical coin toss (one governed by the laws of quantum mechanics), it would tell you that the coin was 50 percent heads and 50 percent tails while it tumbled through the air and landed in the palm of your hand. But as soon as you uncover your wrist, you find you have either a head or a tail—not both. Quantum mechanics does not, at least in its present form, tell you how the 50–50 "superposition" of heads and tails suddenly transformed into 100 percent heads or 100 percent tails. Unfortunately, the equations of quantum mechanics, like those of mechanics and electromagnetism, are entirely time-reversible, so there is no reason why this whole story should not work backward. But it is safe to say that none of us have ever seen a head or tail dissolve into a nebulous combination of both. Although theory has failed, nature has again provided us with a one-way street. The third arrow of time.

The fourth arrow is given to us by one of the most minute and puzzling phenomena of nature: the decay of the elementary particle known as the neutral kaon. Like many other unstable particles, the kaon decays with a characteristic half-life into yet lighter particles. Now, according to a famous theorem from quantum mechanics called

CPT (for "charge-parity-time"), if we take our world, change all the matter into antimatter, reflect it in a mirror, and run time backward, the result should be a world that is indistinguishable from the one we started with. Furthermore, until 1964 it had always been assumed that particle decay proceeded at the same rate whether time ran forward or backward. In 1964 many illusions were shattered when Val Fitch and James Cronin of Princeton University constructed an experiment to observe kaon decay and discovered that the system, transformed into antimatter and reflected in a mirror, decayed at a rate *different* from the original.

Note that the Princeton group dealt with only two of the CPT operations: they changed C (matter-to-antimatter) and P (reflection-in-mirror). Even so, the implications were staggering. If CPT is correct—and there are myriad reasons to think it is—then, under time-reversal alone, the kaon decay rate must change. This is slightly tricky. The matter-to-antimatter transformation followed by a reflection altered the decay rate, so reversing time must *also* change the decay rate in order that the combination of all three operations does nothing. (For a close analogy, take the product of three numbers $CPT = Z$, with Z a constant; for example, $2 \times 3 \times 4 = 24$. If you change C and P to 1 and 2, then you must change T to 12 in order to keep the product Z equal to 24.)

So, here is the neutral kaon that decays at different rates depending on whether time runs forward or backward.

Nature is being very subtle here. Only in the neutral kaon does she reveal the fourth arrow of time. Uranium, thorium, radon . . . all decay at rates independent of the direction of time. Is the kaon's behavior an accident, a minor welding defect in God's construction of the universe? Physicists do not like accidents. As E.C.G. Sudarshan, inventor of tachyons and master of metaphor, has been known to say: "It's like a grain of sand in an oyster—small, but you can't ignore it." The scientific community has not ignored it: Fitch and Cronin were awarded the 1980 Nobel Prize in physics.

If the first four arrows of time are to be found in some of the smallest phenomena of nature and in daily experience, then the last three are to be found in some of nature's grandest phenomena and in objects that have not yet been discovered. And if most physicists would agree on the first four, the last three are subject to nothing but controversy.

Arrow five arises from the distinction between black holes and white holes. Roger Penrose may have introduced white holes to the world by projecting a slide of a black hole upside down, but ironically, he doesn't believe they exist. A white hole, at least to most cosmologists, is taken to be a black hole run backward in time. (A spacetime diagram has space along one axis and time along the other; since time was along the vertical axis in Penrose's slide, inverting the diagram reversed direction of time.) If the universe did not distinguish between forward and backward time, you would expect to see even more white holes than black. But while most astronomers are convinced that black holes exist, not one has yet to observe a white hole. (Cynics would argue that even black holes are hypothetical and this entire discussion meaningless, but cosmologists tend to ignore such unimaginative and feeble obstructions.)

While to physicists black holes are elegant and simple objects, their opposites are troublesome and somewhat incomprehensible. A black hole is formed when a very massive star collapses into an object so dense that even light cannot escape the intense gravitational field. The black hole then sits there quietly forever. If we run the picture backward, we have a white hole that sits quietly since the beginning of time and suddenly, without warning, explodes into ordinary matter. Penrose considers this rather unsatisfyingly unpredictable behavior, a feature made worse by the white hole's singularity.

The singularity of a black hole lies at its center and is the point where all the collapsed matter has fallen, where the density is infinite, and where the known laws of physics break down altogether. Since light cannot escape the black hole, its singularity is hidden to the external world. But everything escapes the white hole, light included, so its singularity is highly visible to even the most casual spacefarer. And because physics has gone haywire at the singularity, there time may go forward or backward; televisions may pop into existence; so, too, may politicians. Anything is possible. Penrose finds such behavior so implausible that he has proposed what we might call the "Black Power Hypothesis," which forbids white holes altogether. If Penrose is correct, then nature provides us with another arrow of time by allowing black holes but not their time-reversed counterparts.

Penrose's longtime collaborator, Stephen Hawking of Cambridge University, disagrees. In 1974 Hawking shook the world of cosmology with his remarkable discovery that black holes are not truly black but radiate energy like any hot body. He proposes that both black holes and white holes radiate in such a way that they are indistinguishable. In that case a fifth arrow of time does not exist. Penrose has counterattacked with arguments of his own, but Don Page of Penn State, Hawking's frequent collaborator and amanuensis, has sided with Hawking, and it is fair to say that at the moment the issue remains unresolved.

Penrose and Hawking have also had disagreements over the sixth arrow of time—the expansion of the universe itself. At present, the universe is cooling and galaxies are receding from each other, sometimes at velocities approaching that of light. In a time-reversed universe galaxies would rush at each other headlong and collide, and instead of cooling down, the universe would heat up until we all fried in a bath of radiation, at which point life would necessarily cease. Quite a different picture from what we actually do see.

Everyone except perhaps creationists agrees that the universe is expanding today. The arguments start when we think about tomorrow. Many cosmologists, notably John Wheeler at the University of Texas, Austin, champion the view that one day the universe will halt its expansion and recollapse, ending its life in a Big Crunch. Will time begin to run backward at the turnaround? Will broken watches reassemble themselves into working timepieces? Will the dead rise again?

Twenty years ago such a point of view was fashionable, but has now fallen out of favor. It is easy to see why. If in an expanding universe time runs forward and in a collapsing universe time runs backward, then presumably at the moment of recollapse time runs in both directions at once, with watches shattering and assembling simultaneously.

Confused? Fine. Increase your confusion by trying to imagine life in such a universe. Though to us superman would appear to fly backward, trees ungrow, and nuclear arsenals diminish, to the "retropeople" who inhabit this time-reversed world, life would plod on mundanely forward. After all, their arrow of time is pointed in the opposite direction to ours, so the universe still appears to expand and watches still shatter. Superman behaves sensibly and trees grow. Yet because their past is our future, and vice versa, communication with these retrofriends would surely be impossible. If he or she broadcast to you a film of a mountain forming, you would see it as eroding, and, like the U.S. and Soviet Star Wars negotiators, come to completely incompatible conclusions about what it all means. Such paradoxes have led Penrose—and most cosmologists—to reject the idea that time runs backward in a collapsing universe.

Until recently, however, Stephen Hawking argued that such a reversal was not unreasonable. Considering the state of the entire universe, he concluded that it should be highly ordered when it is small. So if the universe recollapses, it must become small and highly ordered again. To accomplish this, time must begin to run backward at the turnaround.

Apparently Hawking has changed his mind. Don Page reflects, "I think I managed to convince Stephen that time could still run forward in a collapsing universe and yet satisfy his initial hypothesis." So perhaps Penrose and Hawking are on their way to reconciliation.

You may well wonder what the sixth arrow—the expansion of the universe—has to do with self-assembling watches and retropeople. Are we not confusing two different arrows of time? Exactly. The very paradoxes already mentioned suggest that they have *nothing* to do with each other. The impossibility of retropeople and de-smashed watches are examples of the seventh arrow of time. Seven is a magic number and it is appropriate that the seventh arrow of time has always been considered the most mysterious and difficult to understand; it is the universal increase of entropy.

In terms of confusion-to-understanding ratio, probably no concept in physics rates higher—or has caused more headaches—than entropy. One recent book, *Evolution as Entropy,* equates a process to a quantity, and another, Jeremy Rifkin's *Entropy,* reduces knowledge to ignorance. Entropy is a quantity that appears alongside energy in the equations of thermodynamics, that branch of physics dealing with the transfer of heat in any physical process, such as the burning of gasoline in a car engine or the connecting of a battery to a light bulb. We all learn in school that the total amount of energy in these processes remains the same—it just changes form, from the chemical energy in the gasoline to the mechanical energy in the wheels of the car. This is the first law of thermodynamics: energy is conserved.

Unlike energy, entropy is actually created in most physical processes and tends to a maximum. The statement that entropy always increases toward a maximum is the famous second law of thermodynamics. The easiest way to think about entropy is as a measure of disorder in a system. A watch in hand is a highly ordered system. When it is smashed, scattered all over the floor, it is a highly disordered system. Entropy has increased. When you burn gasoline, some energy goes not into the wheels of the car but into increasing the random motion of molecules in the engine block. This random motion is measured as waste heat—entropy has increased.

You do not see watches auto-assemble, nor the dead rise from their graves; the second law of thermodynamics prevents this and therefore provides the seventh arrow of time. In the universe at large, entropy also tends to a maximum, which means it probably started out from a minimum. Hawking's highly ordered small universe is in a low-entropy state, and his suggestion that the entropy arrow does reverse if the universe recollapses would require that the seventh arrow be tied to the sixth. Unfortunately, it is far from trivial to decide what a minimum entropy state for the entire universe is or means. Penrose postulates that the Big Bang itself was of a very particular type—very smooth, with the lowest possible entropy. Only in this way can entropy now be increasing.

Here, Penrose has appealed to an argument popular among cosmologists to explain the entropy increase in any system. He has chosen the *boundary conditions*—in this case, the Big Bang itself—to be of the type needed to give him the behavior he requires. Although the appeal may be popular, it does invite the immediate question: why did the universe start off in a low entropy state?

Knock on any physicist's door with this conundrum and you will be met with a bewildered shrug; indeed, the last arrow of time remains the most puzzling of the seven and the most controversial. Since French scientists developed thermodynamics about 150 years ago and introduced entropy to physics, no one has given a generally accepted reason why this measure of disorder should always increase. Whereas the other arrows of time may be unexplained in terms of current knowledge, the seventh arrow is firmly rooted in an accepted branch of physics and even merits its own law of nature—the second law of thermodynamics. And still no one understands it. The dilemma is easy enough to see. Take a liter of gasoline and burn it. According to thermodynamics, entropy increases irreversibly—the seventh arrow says there is no way you can run the clock backward and reconstitute your liter of gasoline. But now look at the same liter under a microscope. Each atom obeys Newton's laws and so cannot be subject to an arrow of time. Here we have two great pillars of physics, thermodynamics and mechanics, which evidently rest on totally incompatible principles! Like the church, physics is split into two domains—the reversible and the irreversible. And although many have tried, no one has succeeded in mending the schism.

The great nineteenth-century Austrian physicist Ludwig Boltzmann thought he derived the law of entropy increase from Newtonian mechanics by averaging the behavior of large numbers of particles and predicting their most likely or probable behavior—which turned out to be an increase in entropy. This strikes us as somewhat suspicious, for reasons already given: somehow, from Newton's theory, which displays no arrow of time, you have managed to come up with a theory in which time's arrow is manifest. Indeed, almost as soon as Boltzmann's paper appeared in 1872, critics pounced. Boltzmann's "H-theorem," as it is called, was shown to subtly assume the very conclusions it attempted to prove, and so could not demonstrate the existence of time's arrow. It is generally thought that the setback contributed to Boltzmann's suicide in 1906. On his grave is etched his famous equation relating entropy to probability: $S = k \log W$.

Although Boltzmann's proof was flawed, his viewpoint is still echoed today by many leading physicists: only when you have a large number of particles does entropy increase from low to high. In other words, entropy is a statistical concept and, statistically, it is vastly more probable for entropy to increase than decrease.

This is a very mysterious statement. Imagine two demons, one, called Ludwig, who is very nearsighted and the other, called Isaac, who is very sharp-sighted. Because Ludwig's poor vision blurs the paths of individual particles, he can talk only about statistical averages, and concludes that entropy increases. But Isaac, whose vision is so sharp that he can see the trajectories of each particle, says that they all obey the time-reversible laws of mechanics, and so entropy cannot increase. Not many demons would care to admit that the world is so subjective. You might as well claim that apples fall down for some of us but up for the rest.

The subtle point here is that *the entropy arrow arises from the use of probabilities to predict the future and not the past, not from any intrinsic arrow of time in the equations of mechanics.* (See "The Evolution of Entropy" in *Science à la Mode* for a full discussion of this issue.)

Ilya Prigogine is a Russian-born Belgian who in his youth was trained as a concert pianist; in 1977 he received the Nobel Prize for his contributions to nonequilibrium thermodynamics—a subject that relies on entropy as its central concept. Dividing his time between the Free University of Brussels (free, that is, from the church) and the Center for Statistical Mechanics at the University of Texas in Austin, Prigogine has strong views on the arrow of time. His early artistic training instilled in him the inevitability of evolution: "A piece of music run backward is a different piece of music." He rejects Einstein's position that time is an illusion, as well as Boltzmann's statistical interpretation of entropy. For Prigogine, irreversibility of nature is so pervasive that it must be built in at a very fundamental level.

Prigogine champions the view that there exist "intrinsically random systems" that display an increase in entropy regardless of how few particles are present. The theory of "intrinsically random systems" finds its roots in the study of mathematical chaos, which has become a large scientific industry over the last ten years. A chaotic system may consist of only three particles, but its behavior is so complicated that it rapidly becomes unpredictable.

One of Prigogine's younger colleagues at Texas, Tomio Petrosky, has been developing computer models to illustrate chaotic behavior. He considers a very simple system: the planet Jupiter in orbit around the sun. Then he sends in a comet from a great distance. Now, certain real comets have been captured by the sun and appear regularly—for example, the orbital period of Halley's comet is about 76 years. Other comets swoop in once and never return. In his computer simulations, Petrosky launches comets on "nearly parabolic" orbits—orbits that lie just between capture and flyby. If only the sun were present, any sophomore physics major could calculate whether the comet would be captured or not. But in the presence of Jupiter, Petrosky finds it impossible to predict how many times the comet will circle the sun before flying off again into deep space. For example, when he programs the machine to follow a particular comet's trajectory with six-digit accuracy, he computes 757 orbits. When he increases the precision to seven digits, he gets 38 orbits; at eight digits, 236; at nine, 44; at ten, 12; at eleven, 157, and so on. The comet's trajectory is not changing, only the precision by which it is computed. And the result is totally random. To find the "true" answer, you would need *infinite* precision.

"This is why for some dynamical systems it does not make sense to talk about Newtonian trajectories," says Prigogine. "In these cases they are an idealization that does not exist in the real world." In other words, the demon Isaac would require infinitely sharp eyesight to follow Petrosky's comet. But since infinite information is lacking and the smallest bit of ignorance leads to totally unpredictable behavior, we cannot reverse the clock to bring the system back to its initial position. Petrosky's comet, unlike Newton's, forgets its history and becomes nondeterministic and time-irreversible. But all is not lost. Though Petrosky cannot predict the behavior of an individual comet, he does find that he can predict the average behavior of comets. So here we have recovered something of Boltzmann's dream—chaos, on average, is predictable. Entropy tends to a maximum.

Is Petrosky's model relevant to the real world? Perhaps. Petrosky tells about a seminar he gave in which he explained that many comets launched on similar trajectories would nevertheless have wildly different orbits, some with short periods, others with long, some captures, some flybys—the same sort of diversity exhibited by real comets. He speculated that comets come from a large cloud beyond the solar system and fall toward the sun on their chaotic trajectories. After his talk, a well-known professor, Roman Smoluchowski, explained that Petrosky had independently stumbled upon the Oort cloud, long believed by many astronomers to be the source of the solar system's comets.

Another member of Prigogine's research team, Dilip Kondepudi, is somewhat independent-minded on the subject. "Prigogine's program has been demonstrated for a few specific cases. A general theory is still being developed." Kondepudi goes on to say that "for most simple physical systems, the region in which chaotic behavior occurs is small. For this region, quantities characterizing irreversibility can be derived—as Tomio has for the systems he studied."

Prigogine himself is optimistic. "We are living at a very interesting moment. We can no longer say that irreversibility is due to an approximation or lack of knowledge, and we must now decide what sort of microscopic laws lead to irreversibility."

But Prigogine's ideas tend not to be accepted by the physics community at large. Cosmologists in particular scoff at Prigogine's approach, nod off to sleep in his seminars, and are frequently heard to exclaim "Nonsense!"—the first word in any physicist's vocabulary. But upon closer interrogation one finds that most cosmologists have not taken the time to learn Prigogine's theory. And vice versa.

When one travels between the worlds of cosmology and thermodynamics, one hears two languages spoken. Cosmologists still tend to consider entropy a statistical quantity that arises from lack of knowledge, and whose perpetual increase can be explained by assuming the universe started off in the lowest entropy state possible. Don Page frequently speaks of "coarse-grained entropy," the scientist's term for entropy as an average over ignorance. And contrast the words of Prigogine "we must now decide what sort of microscopic laws lead to irreversibility" with that of cosmologist Paul Davies: "It is remarkable that all the important aspects of time [irreversibility] encountered in the different major topics of physical science may be traced back to the creation or end of the universe."

Neither thermodynamicists nor cosmologists have satisfactorily explained time's seven arrows. It may well be that many, if not all, are linked in ways not yet clearly understood. For today, the seven arrows of time remain a mystery. Someday the realms of the reversible and the irreversible will be joined together, the schism between the followers of Newton and the followers of Boltzmann mended, and the riddle of time resolved. The day may be far off. But let us not lose hope. There is plenty of time.

CHAPTER 4

The Measure of All Things

WHEN a medieval theologian hurrying to matins stepped outside the cloister on a cold winter's night, perhaps he sometimes paused, gathered his cloak about him, and gazed heavenward. If he had not ruined his vision over long years of copying manuscripts (eyeglasses being scarce), he might catch sight of Mars or Jupiter, each of which revolved around the earth on a crystal sphere accompanied by high angels, the Dominations and Virtues. If the moon were out, pushed by lower angels on its own crystal globe, its light would obscure many of the few thousand stars visible to a nearsighted monk. These stars were fixed to yet another sphere, farthest from earth and nearest to God, guided by the Cherubim. A medieval theologian knew this hierarchy of spheres and angels to be correct because, according to Aristotle's Principle of Continuity, nature passes gradually from the least perfect—the contaminated and impure earth—upward to the most perfect—God. Hell was the center of the universe.

Today, when an astronomer walks from his mountaintop dormitory to the telescope dome and pauses to gaze upward at the night sky, he sees a very different universe. He knows that the earth revolves around our sun, which itself lies near the edge of a galaxy containing 100 billion other suns. And our galaxy is itself merely one of a billion that inhabit the observable universe. We became very insignificant indeed when Copernicus shifted hell away from earth.

Aristotle's Principle of Continuity and others like it have thus been supplanted by the Copernican Principle, which holds that one part of the universe is no more privileged than any other. As much as Aristotle's Principle of Continuity was revered by ancient thinkers, so is the Copernican Principle rarely brooked by modern minds. For example, astronomers Sebastian von Hoerner, formerly of the National Radio Astronomy Observatory, and Carl Sagan frequently speak of the Principle of Mediocrity (a modern version of Lovejoy's Principle of Plentitude), which is just a special case of the Copernican Principle. The Principle of Mediocrity states that life on earth is not special; therefore the galaxy should be filled with intelligent civilizations engaged in equally intelligent conversation. (Physicist Freeman Dyson of Princeton's Institute for Advanced Study refers to this as the Philosophical Discourse Dogma.) Thus the widespread belief that extraterrestrial life is commonplace is really a result of the Copernican Principle.

Such non-geocentric reasoning can be extended to construct models of the universe itself. If you believe that no one place in the universe is different from any other place, then the cosmos must be *homogeneous,* which means that all points are indistinguishable from each other. Einstein's original cosmological model and most subsequent models studied to the present day are indeed homogeneous. Usually cosmologists go further and assume that the universe looks the same *in all directions,* in which case the model is said to be not only homogeneous but isotropic. The statement that the universe is homogeneous and isotropic is known as the Cosmological Principle. The familiar standard Big Bang theory is homogeneous and isotropic and is consequently a direct descendant of Copernicus.

The Cosmological Principle can be extended: the universe should not only look the same at every point and in every direction, but should also be identical at all *times.* Such a position is known as the Perfect Cosmological Principle. But in the Big Bang universe the distance between galaxies is always increasing, and so such a model necessarily violates the Perfect Cosmological Principle. Instead, the Perfect Cosmological Principle leads to the Steady State theory, which postulates that as the universe expands, new galaxies are continually created to fill the gaps left by the old ones. In this way, the universe looks the same whether it is one billion years old or ten or one hundred.

However, the Steady State theory is wrong. Twenty years ago Penzias and Wilson at Bell Labs detected a spurious signal in their radio telescope which at the time was explained by Robert Dicke and his colleagues at Princeton University (although it had been predicted years earlier by Alpher and Herman). It was the now-famous cosmic microwave background radiation, radiation left over from the primordial fireball itself, ten billion degrees at one second after the Big Bang, today only 2.7°K after the universe has expanded and cooled for ten billion years. Since the cosmic microwave background radiation indicates that the universe was hotter in the past than it is at present, this automatically rules out the Steady State theory, because the Perfect Cosmological Principle requires that the temperature *then* be the same as it is *now.* The Copernican Principle failed when it was extended to evolution in time.

There is a moral to my tale. Physics is filled with principles. In addition to the Copernican Principle, the Principle of Mediocrity, the Cosmological and Perfect Cosmological Principles, there is Mach's

Principle, which led Einstein to begin work on relativity. But the general theory of relativity, based on Einstein's own Principle of Equivalence, turned out finally to violate Mach's Principle—which had originally inspired it. There is also Heisenberg's Uncertainty Principle, the quantum-mechanical version of "You can't have your cake and eat it too." Another favorite is the Principle of Simplicity (formerly called Ockham's Razor), which says always construct the simplest possible theory. Closely allied with the Principle of Simplicity is the Principle of Beauty, which tells us that the most beautiful theory is the correct one. Now a quiz: would you use the Principle of Beauty to construct Big Bang or Steady State?

There is no answer. Thus: beware the principles of physics. They are not always correct, nor are they always compatible.

And now comes another principle, discussed at cocktail parties, debated in the *New York Review of Books*, glorified in John Updike's latest novel *Roger's Version*: the *Anthropic Principle*. Or as the sophist Protagoras said 2,500 years ago, "Man is the measure of all things."

The term "Anthropic Principle" was coined in 1974 by cosmologist Brandon Carter, then at Cambridge University, during a lecture for the International Astronomical Union. He saw the Anthropic Principle as a foil against taking the nonevolving Copernican Principle to the (absurd) extremes of the Steady State theory and the Philosophical Discourse Dogma. Carter's point was in accord with the fundamental rule of physics that you must always take into account the properties of your experimental apparatus, in this case, us: "What we can expect to observe [in the universe] must be restricted by the conditions necessary for our presence as observers. (Although our situation is not necessarily *central*, it is privileged to some extent.)" In other words, the conditions we observe in the universe are those necessary to give rise to intelligent life, otherwise we wouldn't be here to observe them.

This "biological selection effect," as Carter termed it, is best illustrated by returning to a 1961 paper by Robert Dicke, in which he made the first fully developed anthropic argument. To simplify somewhat, Dicke was trying to answer the following question: why do we observe the universe to be approximately ten billion years old? (see box.) A Copernican response might be that one age of the universe is just as good as any other, and it is merely a *coincidence* that we happen to observe the universe's age to be roughly ten billion years. Dicke's answer, on the other hand, was very different. He reasoned that the universe must be at least old enough to have generated elements heavier than hydrogen because "it is well known that

The Mystery of the Large Numbers

Of all the mysteries that the universe confronts us with, few have inspired more speculation than the riddle of the Large Numbers, a riddle that is inextricably linked to the names of Eddington, Dirac, and Dicke.

In physics, "dimensionless numbers" are of particular importance. A dimensionless or "pure" number is one in which all the units, such as centimeters, grams, or seconds, have canceled. The number 1 is dimensionless; the number 5 mph is not. Dimensionless numbers are important because they are usually formed from the ratio of two like quantities and thus measure the size of one thing relative to another. For example, a car can easily travel at 30 meters per second. The speed of light is 3×10^8 meters per second. Their ratio is (30 mps)/(3×10^8 mps) $= 1 \times 10^{-8}$. The fact that this number is so small tells us that Einstein's relativity is not important while traveling in a car. Were the ratio near 1, this would mean that the car is traveling at nearly the speed of light, and relativistic effects should be taken into account.

In physics, most naturally occurring dimensionless numbers turn out to be near one. The ratio of the proton mass to the neutron mass is almost exactly one. The ratio of the proton mass to the electron mass is almost 2000, which may seem large until you consider that the number of protons in the observable universe is

$$N_p \sim 10^{78}, \qquad (1)$$

where $\sim$ means equal to within a factor of ten or so.

In 1923 Arthur Eddington noticed a peculiar coincidence involving the gravitational constant G, Planck's constant h, the speed of light c, and the mass of the proton m_p. He formed what is usually called the "dimensionless gravitational coupling constant":

$$hc/(Gm_p^2) \sim 10^{39}. \qquad (2)$$

Note that to the accuracy of the $\sim$, this number is just the square root of (1). There is no obvious reason why this collection of fundamental constants should have anything to do with the number of protons in the universe, so you can dismiss this approximate square root between (1) and (2) as a coincidence; Eddington invented a theory to explain it.

Paul Dirac noticed a similar coincidence. It turns out that the ratio of the electric force between a proton and an electron to the gravitational force between the same two particles is very nearly equal to the ratio of the age of the universe to the time it takes light to cross an atom. In symbols,

$$\frac{\text{(electric force between proton and electron)}}{\text{(gravitational force between proton and electron)}}$$

$$= e^2/(Gm_pm_e) \sim 2.3 \times 10^{39} \qquad (3)$$

and

$$\frac{\text{(age of universe)}}{\text{(time for light to cross atom)}}$$

$$= t_u/(e^2/m_ec^3) \sim 6 \times 10^{39}. \qquad (4)$$

Here we have introduced the age of the universe t_u and mass of the electron m_e. Now we have something really remarkable. There is certainly no reason that the age of the universe should have anything to do with this collection of fundamental constants, much less that these two enormous numbers should be so nearly equal to one another. Another coincidence? Perhaps. Dirac assumed it was not, and that for mysterious reasons (3) and (4) should always be *equal* to one another. If you set them equal to each other and solve for the age of the universe, the "coincidence" can be written

$$t_u \sim (1/G) \times e^4/(m_pm_e^2c^3). \qquad (5)$$

Notice that everything in this equation except t_u—which naturally increases—is normally considered a constant. Something has got to give as the universe gets older. Dirac reasoned that all the constants on the right except G govern atomic phenomena, while the large-scale structure of the universe is governed by G itself. Thus, assume all the atomic constants are in fact constant but let G vary. Then (5) immediately tells you that G is proportional to $1/t_u$. The gravitational constant decreases as the universe gets older. Furthermore, if you really believe that (1) should always be the square of (2), you also get that N_p increases as t_u^2—in other words, matter is created. This "Variable-G" theory is called the Dirac theory of gravity, which has since been ruled out by experimental evidence.

But Robert Dicke had another explanation for the "coincidence" (5). He calculated from basic nuclear physics that the main-sequence lifetime of a star should *also* be roughly equal to the collection of constants on the right-hand side of (5). If you assume that the universe must be at least as old as stars for life to exist, then it is no coincidence that we see (5) satisfied, since otherwise we wouldn't be here. This is one of the first modern uses of the Anthropic Principle.

carbon is required to make physicists" (at least physicists as we know them). Carbon in fact has not existed since the Big Bang. Rather, it is cooked near the center of stars, then scattered throughout space when and if the star explodes in a supernova. The cooking time depends on the mass of the star but averages a billion years or so. To observe a universe younger than the lifetime of a star is excluded, since the very elements of which we are composed would not yet exist. We are truly starchildren.[1]

Furthermore, if the universe were much older than ten billion years, it is almost certain that most stars, would have disrupted or collapsed into white dwarfs, neutron stars, and black holes, rendering our type of life impossible once more. This type of reasoning led Dicke to conclude that the fact we observe the universe to be roughly ten billion years old is not a Copernican accident but a necessary result of biological selection. The universe's observed age "is limited by the criteria for the existence of physicists."

Although Dicke gave no name to his argument, thirteen years later Carter would term it an example of the Weak Anthropic Principle. That is, while the Copernican Principle requires that all locations and physical quantities be equally likely or unprivileged, the Weak Anthropic Principle (WAP) says no: the observed values of all physical quantities in the universe are restricted, or privileged, by the requirement that they be compatible with the existence of carbon-based *Homo sapiens*.

I find little to quarrel with in the WAP; as Carter implied, it merely extends to humans the usual rule of taking into account the properties of our measuring apparatus when we do experiments. As commonsensical as this may be, however, the WAP can in theory be used to make predictions, and some proponents would argue that it is already doing so. Let me give some examples of missed predictions, ignored predictions, and conceivably testable predictions.

An example of a missed prediction: Dicke could have used the WAP against the Steady State theory. He required that the age of the universe be about equal to that of a star. Astronomers cannot actually measure the age of the universe directly; instead, they measure a quantity called the Hubble "constant" (denoted by H). In the Big Bang theory, the age of the universe is approximately the inverse of

[1] The numbers used here are very rough. For those who insist on closer agreement between the age of a star and the age of the universe, to the billion years to which we are entitled add a few more for planet formation and the time necessary to evolve intelligent life. So if you believe this argument, we could not observe the universe to be much younger than it actually is.

the Hubble constant ($1/H$), which shows that H is not really constant but gets smaller as the universe gets older. In the Steady State theory, by contrast, H really is constant, and furthermore bears no relationship to the age of the universe. (Strictly speaking, the age of a Steady State universe is infinite; after all, it *is* steady state.) Thus, in a Steady State model there is no reason for ($1/H$) to equal the age of a star. If it does, it is merely a coincidence. By using the Anthropic Principle, you reject the coincidence and decide in favor of the Big Bang model. Dicke might have done this, though it apparently did not occur to him. And, in fact, today we know that H has gotten much smaller as the universe evolves—at the instant of the Big Bang, H was actually infinite.

On another occasion the WAP was used perhaps more convincingly, but ignored. During the nineteenth century a great debate raged over the correctness of Darwinian evolution (a debate I fear will never end). The physicist Lord Kelvin opposed Darwin, and as an argument against him calculated that the sun, producing energy by known processes, could never shine long enough for natural selection to have produced man. Kelvin's argument was so obviously correct that it was widely used to discredit evolutionary biologists. One lone voice dissented. The American geologist Thomas Chamberlain assumed that evolution had taken place and concluded that the sun must burn by a new source of energy. "Certainly no careful chemist," he wrote, "would affirm that . . . there may not be locked up in [atoms] energies of the first order of magnitude . . . Nor would they probably be prepared to affirm or deny that the extraordinary conditions which reside at the center of the sun may not set free a portion of this energy." These words were written in 1899, over half a century before the explosion of the first hydrogen bomb.

One of the most spectacular anthropic predictions was made by Fred Hoyle in 1954, but has never received any publicity. At the time, he was attempting to understand the nuclear reactions that take place in the sun. On the basis of the Anthropic Principle he argued that a certain reaction producing carbon must take place "resonantly," or with extreme efficiency; otherwise not enough carbon would be produced in stars to support life. At the same time a different reaction, transmuting carbon into oxygen, must not take place resonantly, or else all the carbon formed by the first reaction would be depleted. The resonance making the first reaction efficient was shortly discovered exactly where Hoyle predicted it; the second reaction was found to be nonresonant.

More recently the WAP has been extended to the debate over extraterrestrial life. Once upon a time I asked the physicist John

Wheeler for his views on the subject, and he replied cryptically, "Less is more." After some years I began to understand what he meant. Wheeler was using Dicke's conclusion as a foil against the Principle of Mediocrity, the idea that the universe is filled with intelligent extraterrestrial life. If the universe must be at least a billion or ten billion years old for us to observe it, then because it is expanding at nearly the speed of light, it must be nearly ten billion light years across. Supporters of the Principle of Mediocrity gaze into this vastness and are certain that hidden amidst the million trillion suns are myriad civilizations. We have all shared this feeling. But from the Dicke-Wheeler anthropic viewpoint, the universe must be as big as it is merely to give rise to one intelligent civilization—us. Additional civilizations would, according to Wheeler, be wasteful. "Less is more."

In 1983 Brandon Carter made this sort of argument more precise and produced a mathematical formula known as Carter's inequality. It assumes that there are n very improbable steps toward intelligent life. Next he predicts (with n) the average time between the emergence of an intelligent species and the time that it will end because, say, its sun burned out. The important point is that Carter assumed life was improbable. If extraterrestrial life is discovered with any frequency, then his formula is incorrect. So here is a prediction of the WAP that is not only testable but is now being tested by the various searches for ETI.

Unfortunately, one does have to know n, the number of improbable steps toward intelligent life, and here contemporary biology is unable to help. Nevertheless, we might still get an idea of whether Carter's formula is plausible by looking at various traits considered crucial to the evolution of earthbound life. If these traits are determined to have arisen independently in each species, we would judge the trait probable and Carter's formula incorrect. If these traits are found to have originated from a common gene sequence, we would judge the step improbable, and this would lend support to Carter's inequality. Although evolutionists are working on this problem, the lack of consensus is absolute.

If the Weak Anthropic Principle can be accepted as commonsensical, the same is not true for what has become known as the Strong Anthropic Principle (SAP). As with the WAP, the SAP is found worded in various confusing ways. Carter originally formulated it thusly: "The universe *must* be such as to admit the creation of observers within it at some stage." One might well respond that this is trivial: the universe *has* given rise to observers. Be that as it may, most scientists interpret the SAP to mean that the universe *must* be exactly as we know it or life would not exist; "conversely," if life did not exist, neither would the universe.

An obvious difficulty with the SAP is that (apparently) only one universe lies at our disposal for the purposes of experimentation. We cannot claim with any certainty that life would or would not exist in a universe in which, say, energy were not conserved. Even conceiving of such a universe is difficult. (My own suspicion is that life would be *very* possible in such a universe; almost everything would.) What is easier to conceive of, though at the moment equally far from the grasp of experimentalists, is a universe in which the fundamental constants differ from their known values.

In the canon of numbers important to a physicist, the constants of nature assume the most sacred position. Of these, the most exalted are Newton's gravitational constant G, Planck's constant h, the speed of light c, and Boltzmann's constant k. The numerical value of G determines the strength of the force of gravity.[2] Were G larger, baseballs would fall to the ground faster than we observe them to do so, and the number of home runs in any season would drastically decrease. In a similar way, h governs the size of quantum effects. We know that because of the presence of h, electrons in atoms are permitted to orbit their nuclei only at discrete or quantized frequencies. Were h roughly 10^{34} times larger than it actually is—and if we make the fantastical assumption that stable matter would still exist—riding a bicycle would become impossible, because the wheel would be allowed to turn only at 0 rpm, then at 1 rpm, 2 rpm, etc.—and at no speed in between.

More serious effects would result if God tinkered with the values of the other natural constants. For instance, astronomers have found that about 25 percent of the universe's mass is concentrated in the element helium. Most cosmologists are certain that this helium was formed during the famous "first three minutes" after the Big Bang, when the universe was hot enough to allow nuclear reactions to take place. The actual amount of helium formed depends on many factors—principally, on the rate at which the universe is expanding versus the rate of the nuclear reactions (see box). The expansion rate of the universe, in turn, depends on G (among other things), while some of the crucial reactions depend on the "weak interaction cou-

[2] Physicists will properly object that I should use the so-called dimensionless coupling constants in this discussion and the sequel. However, for the purposes of exposition I resort to the more familiar dimensioned constants. Physicists are at liberty to convert.

As discussed on page 67, some of the light elements—principally helium and deuterium—are produced in the Big Bang. In the simplest, or standard, model about 25% helium results, the exact amount depending on the value of the constants G and α_w, among other things.

It happens that at the University of Texas, Austin, Richard Matzner and I developed a computer code that we called in its various incarnations Zeus, Odin, Wotan, or just plain God. Its purpose is to calculate Big Bang element formation in nonstandard cosmological models. Although we usually do not touch the values of the constants, it is not difficult to do so, and let the program calculate what happens to helium and the other isotopes. At my request, Richard played God and tinkered with some of the fundamental constants to demonstrate what happens. The following conference call between Richard, Frank Tipler, and myself resulted:

Richard: OK, the standard model run gives 25% helium. Now, if I make α_w about 30 times smaller, we get 90% of the universe in helium. I'm sure if I make α_w smaller yet, we'll get 100% helium, as expected.

Frank: With 90% helium most stars would be helium stars, which burn very fast. So their main-sequence lifetime is not very long, my guess is about a tenth of our sun's main-sequence lifetime or much less than a billion years. That means life wouldn't have much time to evolve before the sun left the main sequence—remember we only came out of the sea half a billion years ago. Also, it would be a very hot star so it would cause a lot of stability problems with our atmosphere—

Tony: You'd blow it away?

F: Well, you'd get a runaway greenhouse effect. According to Michael Hart's work at Trinity College, earth would become like Venus. Liquid water would be impossible and probably life. Of course with 100% of the universe in helium, there can't be any water to begin with.

R: But I can compensate for the effects of lowering α_w by increasing the mass difference between the proton and the neutron. I have a run here where I made the mass difference 6.5 times the usual value and that brought helium back down from 90% to 23%, quite acceptable for life. Of course, such a large neutron-proton mass difference makes deuterium unstable—

T:—so presumably you don't form any helium in the Big Bang. How does this prevent life?

F: Unstable deuterium will also block the p-p chain in stars, since any deuterium you form will quickly disintegrate. This basically stops the production of all heavy elements. No life in that universe.

R: Well, instead of changing the neutron-proton mass difference I can change G. If I reduce G by a factor of 100, helium goes down from 90% to 80%. If I reduce G by a factor of 1000, helium goes down to 74%.

F: A smaller G makes all the stars radiative, not convective. Radiative stars don't shed angular momentum to make planets—

R: Pretty tenuous—

T: Yeah, and this argument assumes that you've kept the mass of the electron constant. What if I compensate for the small G by making the electron mass smaller too? Is there an anthropic argument against that?

F: I'm sure there is, though I can't generate one over the telephone.

R: Now, if you made the electron more massive you'd make molecules smaller and get nuclear fusion at low temperatures—in fact, zero temperature. You'd be back to 100% helium.

T: Any more runs?

R: Yes, you know as we increase G from the standard value, helium increases to a point, but then comes back down. If I increase G by a factor of 1000, I'm over the hump and down to 38% helium, with a lot of neutrons.

T: But they all decay into protons.

F: Doesn't save you. If you increase G, all stars are convective red dwarfs. They allow planets to form but are very cold stars. So you get runaway glaciation and the earth freezes over.

T: Can't you move the earth closer to the sun?

F: You have to look at the entire evolution. Hart finds that if you move earth in even slightly, you go back to runaway greenhouse. It's very delicate. The moral is that you have to consider a whole complex of effects. You can't just assume that because you've gotten away with changing one constant, the road is clear.

R: I agree. Big Bang nucleosynthesis is comparatively simple. Even if you can juggle the constants to produce a reasonable amount of helium, you are almost certain to have screwed up something down the line.

T: In particular, life.

pling constant" α_w, which governs the rate at which neutrons radioactively decay into protons.

It turns out that if α_w were smaller than it actually is, then you would produce 100 percent helium in the Big Bang. A universe of 100 percent helium would almost certainly lack water—which requires hydrogen—and therefore would probably not give rise to our type of intelligent life. By this argument, α_w could not be much smaller than we measure it to be. If it were much larger, then the universe would contain zero helium. While a universe with no helium might be compatible with life, other considerations suggest that a large α_w would prevent supernovae from exploding. And as I have already pointed out, supernovae seem to be essential for life as we know it.

Similar arguments concern *G*. In his 1974 lecture, Carter proposed that if *G* were slightly larger, all stars would be blue giants; if *G* were slightly smaller, they would be red dwarfs. Evidence indicates that planets do not form around blue giants, consequently prohibiting life, while red dwarfs do not produce supernovae, leading to the same lifeless universe.

An obvious objection to this sort of thinking is that perhaps life does not require planets and cosmologists have been unduly anthropocentric in their deliberations. Indeed, for the above discussion to be correct, one would have to show that varying *G* also prohibits life forms such as Fred Hoyle's black cloud, which traveled in interstellar space and was maintained by electromagnetic energy. Otherwise the SAP has failed.

Literally hundreds of such arguments abound in the anthropic literature. The more sanguine anthropocists claim that all the fundamental constants and many of the elementary particle masses are fixed by the SAP. A number have even argued that space has three dimensions because life could not exist otherwise. Nevertheless, for the moment, you are at liberty to accept arguments as reasonable or reject them as wildly metaphysical. But the speculation is far from over. In an attempt to provide a philosophical framework for the SAP, Carter enjoined us to think in terms of an ensemble of universes. Imagine all possible worlds. In the first, *G* may be large, in the second, small. In the third *h* is zero, in the fourth α_w is infinite . . . Only in those universes compatible with life will wayward physicists write articles like these.[3]

[3] Although the notion of an ensemble of universes is used to support arguments for the SAP, strictly speaking SAP posits one and only one universe. If there are truly many worlds to be selected from, it should be the WAP and not the SAP that does the selecting. However, since Carter's original argument related to the SAP, I have chosen to ignore this fine point.

The idea of an ensemble of universes selected by the SAP strikes many as surreal and unnatural. Yet for almost thirty years certain physicists have been convinced that the only logical and self-consistent interpretation of quantum mechanics requires just such an ensemble of universes. It is called the Many Worlds Interpretation of quantum mechanics, first proposed by the late Hugh Everett in 1957 and subsequently championed by Bryce DeWitt and, for a time, John Wheeler.

The origins of the Many Worlds Interpretation lie in Schrödinger's famous cat paradox, whose schizophrenic feline has been the subject of so many popularizations that she has undoubtedly shriveled up and died under the glare of overexposure. Consider a cat imprisoned in a box with a flask of poison gas. A Geiger counter is set to release the gas should it detect the radioactive decay of an atom with a half-life of one hour. By the definition of half-life, this means that as the clock strikes one hour, there is a 50–50 chance that the particle has decayed and released the gas. Interpreting this situation according to the equations of quantum mechanics, we do not say that there is a 50–50 that the cat has been killed, but rather that the cat is in a "superposition" of two states: 50 percent live cat and 50 percent dead cat. (The situation is identical to that discussed in arrow three of the previous chapter.)

Schrödinger rejected this situation as absurd; a cat is either dead or alive, not both. The fact that we observe only live cats or dead cats has led to a long-running debate about when the "superposition" of live and dead cat "collapses" into one or the other. Because quantum mechanics itself does not tell you how this collapse occurs (or in fact whether it occurs), it is often said that the observer, by his action of peering into the box at the end of the hour, actually forces the collapse to take place.

But the number of resolutions to the cat paradox probably exceeds the number of physicists who consider it. Everett took quantum mechanics to its extreme: the theory does not say that the superposition collapses, so assume that it doesn't. Instead, the universe splits along with the observer, and in one branch he sees a live cat and in the other branch his near-replica sees a dead cat. At each experiment such branching occurs until you have 10^n possible worlds, where n is a *very* large number.

Critics object to all of this as untestable metaphysics. This may or may not be so. David Deutsch of Oxford has devised a clever thought experiment that involves a computer with a reversible memory; the computer can remember that it split during a quantum-mechanical experiment. Whether Deutsch's thought experiment has any bearing on reality, I hesitate to say.

The relationship of Schrödinger's cat to the Anthropic Principle is clear. Carter's ensemble of universes becomes the many worlds of the Everett interpretation, and each of the many worlds contains its own values of the fundamental constants. Our world is one in which the values of these constants are compatible with life.

Quantum-mechanical reasoning has led Wheeler to propose an even more radical version of the Anthropic Principle, which he termed the Participatory Anthropic Principle. Let's forget splitting universes and assume that the observer *does* collapse the superposed Schrödinger's cat into one that is fully dead or fully alive. Thus in some sense the observer is necessary to bring reality into focus; before the observer opened the box, reality was a nebulous mixture of dead and live cat. By his participation, the experimenter brought reality alive—or dead. As Wheeler used to say, "No phenomenon is a phenomenon until it is an observed phenomenon."[4]

But the observer himself is a quantum-mechanical system. What collapses *his* wave function into one state or another? His next-door neighbor? Perhaps. And then *his* neighbor . . . ad infinitum. Wheeler has speculated that, in this way, the universe comes into gradual existence as more and more observers are created as midwife. This is the Participatory Anthropic Principle.

The PAP leads directly to the FAP, or Final Anthropic Principle: the universe must not only give rise to life, but once life is created it endures forever, molding the universe to its will. At some time in the future, termed by Teilhard de Chardin the Omega Point, the observers will have brought the universe into full existence. Thus man—or Life—is elevated not only to the measure of all things but to the creator of all things.

When you stand outside your home on a cold winter's night in the late twentieth century and gaze into the vault of heaven, you have a choice of world views: Copernican or Anthropic. Which do you choose? Not surprisingly, the kind of speculation I have been engaging in has drawn fire, more than a little. Wheeler, in particular, has been severely criticized as returning to the solipsism of philosopher George Berkeley, whose name is associated with the idea that if no one hears a tree fall in the forest, then the tree hasn't made a sound.

[4] He later emended this to the less compelling but more accurate "A [quantum] phenomenon is not yet a phenomenon until it has been brought to a close by an irreversible act of amplification."

Several years ago John Barrow and Frank Tipler began a series of papers that triggered a new round of anthropic dogfighting, with both papers and dogfights culminating in the publication of their book, *The Anthropic Cosmological Principle* (Oxford University Press, 1986). The 700-page tome, from which I have drawn a few illustrations, is indisputably the most comprehensive summary of anthropic and related lore ever attempted. Readers are taken on a wide-ranging tour of cosmology, teleology, quantum mechanics, biochemistry, and the ultimate fate of life itself. The work of Barrow and Tipler, as well as the Anthropic Principle in general, has received several large doses of venom from science writer Martin Gardner in the May 8, 1986, *New York Review of Books*, from Stephen Jay Gould writing in *Discover* (March 1983), and from physicist Heinz Pagels in *The Sciences* (March/April 85) (all of which I am sure pleases at least Tipler, who seems drawn to controversy with suicidal enthusiasm). Gardner goes so far as to term the next step after WAP, SAP, PAP, and FAP the CRAP—Completely Ridiculous Anthropic Principle. Whether or not one chooses to accept the Anthropic Principle in any of its variations, I confess to finding many of the arguments brought against it both illogical and misinformed.

Let me begin with Gould's article "The Wisdom of Casey Stengel," since this is simplest. As we have seen, the WAP arguments of Dicke, Wheeler, and Carter indicate that extraterrestrial life may not be so prevalent as one would expect on the basis of the Copernican Principle. Tipler has gone so far to claim that we are the only intelligent civilization in the galaxy. The crucial question, of course, is whether the evolutionary pathway to intelligent life is common or rare. Gould claims that Tipler, in arguing that the path to intelligent life is highly improbable, has misunderstood the views of evolutionary biologists. Evolutionary biologists do not speak with one voice, but I have seen the correspondence between Tipler and Ernst Mayr and George Gaylord Simpson, on whose work Tipler largely bases his case. They state quite unequivocally that Tipler has interpreted them correctly.

Pagels, in his outing against the Anthropic Principle, writes that "Carter's worldview is the product of an anthropocentricism as profound as that which underlay the pre-Copernican view of the universe; the anthropic principle is born of a most provincial outlook on what life is. Its adherents assume that all life must resemble, in broad form at least, life on this planet." He then goes on to produce Hoyle's black cloud as a counterexample to the Anthropic Principle.

This criticism, however, was answered by Carter himself in an article referenced by Pagels. In 1983 Carter said, "If I had known that the term 'anthropic principle' would come to be so widely adopted, I would have been more careful in my original choice of words. The imperfection of this now standard terminology is that it conveys the suggestion that the principle applies only to mankind. However, although this is indeed the case as far as we can apply it ourselves, it remains true that the same self-selection principle would be applicable by any extraterrestrial civilization that may exist."

More serious is Pagel's charge that "unlike other principles of physics, the anthropic principle is not testable. It is all well and good to imagine universes with various gravitational constants and estimate the prevailing physical properties, but there is no way we can actually go to an imaginary universe and test for life. We are stuck with our universe, and powerless to alter its fundamental constants."

There is some justice here; I have already pointed out that, at least as far as the SAP is concerned, only one universe is at our disposal for experimentation. (However, I can vaguely imagine a future civilization being able to change the natural constants of *this* universe.) Nevertheless, Pagels misses the mark when he implies that the "other principles of physics" are testable. The Heisenberg Uncertainty Principle is a precise mathematical statement and testable. Einstein's Principle of Equivalence is also testable. Is the Principle of Beauty, by which most particle physicists swear in their relentless search for symmetry in nature, testable? Not in the sense that you can falsify it. But it has borne great fruit in the Weinberg-Salam theory, for example. The Copernican Principle and the Principle of Simplicity, in changing our world views, have led to relativity and cosmology. Neither are in and of themselves highly testable, and the Anthropic Principle, as demonstrated by Carter's theorem, is certainly the equal of the Copernican Principle in its ability to make predictions.

One can go further. If I used the Strong Anthropic Principle to successfully predict the masses of five subatomic particles necessary for life as we know it but theretofore undiscovered, what physicist would not be impressed? Such a prediction closely parallels Hoyle's prediction of the stellar nuclear reaction rates already discussed; there is no obvious reason why the Anthropic Principle could not be used to predict particle masses. The utility of the principles of physics lies not only in their falsifiability (as Pagels seems to imply) but in their value to serve as signposts that point the way toward more and more successful theories. The Anthropic Principle falls into this category.

On a slightly less philosophical and more scientific level, both Pagels and Gardner dismiss the Anthropic Principle as superfluous in light of such theories as inflation. I believe they have erred here. At best the inflationary universe model is subject to the same criticisms as the Anthropic Principle; at worst it is based on the Anthropic Principle. The inflationary universe was introduced in 1981 by Alan Guth to explain two long-standing cosmological puzzles: why is the universe so isotropic and why is the universe so flat? As I mentioned earlier, an isotropic universe is one that looks the same in all directions, as our universe more or less does. A "flat" universe is one that lies just at the borderline between "closed" and "open," the former being a universe that will eventually halt its expansion and recollapse, the latter being a universe that will expand forever. Our universe also appears to be reasonably flat.

It turns out that the standard Big Bang model cannot answer these two puzzles other than to simply assume that the universe started off flat and isotropic, which strikes many cosmologists as highly improbable. This difficulty led Barry Collins and Stephen Hawking to attempt an anthropic explanation. Basically they argued that in a non-flat (or curved), non-isotropic ("chaotic") universe, galaxies would not form, and consequently life would not arise. Therefore we must observe our universe to be flat and isotropic. (I should mention that the proof is flawed, but that is not the point.)

The inflationary scenario tries to resolve the same puzzles with a more conventional explanation. The process of inflation, during which the universe expands by at least thirty orders of magnitude in the incredibly short time of about 10^{-30} seconds, smooths out any initial irregularities that may have been present and thus isotropizes the universe. The same mechanism makes the universe extraordinarily flat. To slightly oversimplify, inflation takes *any* initially curved universe and flattens it into the one we see. (For a more detailed discussion of inflation, see "Metaflation?" in *Science à la Mode*.)

There is first the technical question of whether the inflationary scenario actually works, but I wish to avoid that debate. Here I want to examine only the logic of the discussion. First, I am always at liberty to assume the universe started off flat and isotropic, in which case inflation itself is superfluous. You may find this improbable and distasteful but, to echo Pagels, we are stuck with our universe and the words "probable" or "improbable" when applied to the origin of an entire universe have little or no meaning. So if I claim that in the beginning the universe was isotropic, you cannot disprove me. This is especially clear if you believe inflation succeeds in isotropizing *any* universe; then there is no way we can decide what the universe was like *before* inflation. Consequently, the questions that inflation attempts to answer can be dismissed as metaphysical. The statement that inflation has actually selected and isotropized a chaotic universe is no more or less testable than the anthropic explanation.

Now look more closely at the flatness problem. Einstein's equations tell us that the universe must eventually become curved again[5] unless it was *exactly* flat to begin with. This is true even if inflation works, except the time for curvature to manifest itself will be unimaginably long. The flatness problem asks: If the universe must become curved again at some point, why does it happen to look so flat to us? Is this a coincidence? Well, barring exact flatness (in which case inflation is unnecessary), if we wait long enough, say 10,000 billion ages of the universe or longer, it will eventually become curvature dominated. Any civilization existing at that time will not notice a flatness problem. You may respond that civilizations will not exist 10,000 billion ages of the universe in the future because all the stars will have burnt out. But then what have you done? You have made an anthropic argument to explain the flatness problem. One leading researcher in inflation, A. D. Linde in Moscow, puts it somewhat differently. The universe begins chaotically; some regions inflate and others do not. Only in those regions that inflate sufficiently to isotropize and flatten the universe will life evolve to discuss the matter.

The debate has brought us to an interesting juncture. The medieval theologian who gazed at the night sky through the eyes of Aristotle and saw angels moving the spheres in harmony has become the modern cosmologist who gazes at the same sky through the eyes of Einstein and sees the hand of God not in angels but in the constants of nature.

We do not seem, on second glance, to have traveled very far. And here lies both the attraction and the fear of the Anthropic Principle. It is not a big step from the Strong version to the Argument from Design. You all know the Argument from Design. It says that the universe was made very precisely, and were it ever-so-slightly different then man would not be here. Therefore Someone must have created it.

Even as I write down the conclusion my pen balks, because as a twentieth-century physicist I know that the last step is a leap of faith, not a matter of logic. Then I reflect. Is it inconceivable that a future civilization will meet God face to face? Will He intentionally reveal

[5] Strictly speaking I should say not "curved" but "curvature dominated," which means that even in the open case, when the universe becomes flatter and flatter as it ages, the curvature terms in the equations dominate and the density is driven away from the critical value. Laymen should ignore the distinction. Specialists will forgive the loose talk.

Himself? Or will we become God? That is, after all, what the Final Anthropic Principle claims. In the face of such speculation I retreat and ask: Are the followers of the Anthropic Principle then attracted to it for scientific reasons or religious reasons? Dale Kohler, the young antagonist in John Updike's novel *Roger's Version*, propounds the naturalist view that science can have something to say about religion. He claims that physicists call themselves atheistic materialists but then try to find all sorts of ways around it, subconsciously injecting God into science. Protagonist Myron Kriegman demolishes his anthropic arguments for the existence of God by using Pagels-like reasoning; God and physics have nothing to say to each other.

Although Kriegman is left with the upper hand, I at least sympathize with Kohler's view of physicists. When confronted with the order and beauty of the universe and the strange coincidences of nature, it is very tempting for a physicist to take the leap of faith from science into religion. I am sure many want to. I only wish they would admit it.

CHAPTER 5

On That Day, When the Earth Is Dissolved in Positrons . . .

THE Old Testament opens with the first great question: Where did we come from? and the New Testament closes with the second: Where are we going? Between Genesis and Apocalypse other questions are incidental. Throughout history these two great riddles have confronted anyone who has found a few moments to put aside the daily task of survival. Incurable romantics, like artists and scientists, are frequent victims of cosmic meditation. Artists have been inspired by our destiny far more often than by our origins. For every master who has recorded Genesis, probably a dozen have chosen to set forth their apocalyptic visions. Yes, there is Michelangelo's *Creation of Adam* on the Sistine ceiling. But there is also his *Last Judgment* above the altar, Giotto's frescoes in the Peruzzi chapel of S. Croce (where for the first time Death carries a scythe), Hubert and Jan van Eyck's Ghent altarpiece, the Brussels tapestries, and Dürer's woodcuts, not to mention the nightmares of Bosch, Goya, Redon, and Landau, as well as innumerable icons, requiem masses, and dirges.

Paradoxically, scientists, who have at their disposal the predictive power of physics and mathematics, have been far more concerned with origins than destiny. This is especially true of cosmology, the branch of physics that deals with the large-scale structure of the universe. While thousands of papers appear annually on the origin of the cosmos, probably fewer than a dozen in the last twenty years have speculated on its ultimate fate.

The paradox is easily explained. Unlike most sciences, cosmology has only one laboratory and one experiment. The laboratory is the universe itself and the experiment has already taken place—more precisely, is taking place—and there is nothing we can do at present to alter the experimental parameters. When we look at distant objects in the sky, such as quasars and Seyfert galaxies, we are peering back in time, because the light from these objects has traveled millions or billions of years to reach us. So it is much easier for cosmologists to study the past than the future.

Technical obstacles are only part of the problem. Scientists genuinely fear predicting the future. They know that a leaf falling from a tree on a wind-blown day is a system whose behavior is too complicated for physics to describe completely. Physics tells you that the leaf will eventually hit the ground, but cannot tell you where. To predict the future of the entire universe is a task conservative scientists would prefer to leave to Madame Sosostris and her bad cold.

Nevertheless, certain reckless physicists have used the crystal ball of science to penetrate the shrouded fate of the universe. The endeavor is perhaps not quite as headstrong as it sounds. The Big Bang models cosmologists use to describe the universe are in fact considerably simpler than a leaf falling on a windy day, and their future behavior is not difficult to determine. The problem lies rather in the opposite direction: the models are so simplified that they may bear no more resemblance to the real universe than a Brancusi bird does to a real bird. Any forecast based on these models will undoubtedly be incorrect in many details. But we can hope that our theories, like the Brancusi bird, are at least correct in the broadest outlines. We can also expect that the laws of physics, valid up to now, will continue to hold indefinitely. Reckless assumptions? If so, then what follows is the purest fantasy. If not, then here is a glimpse of the ultimate apocalypse.

Many of you will already know that according to Einstein's theory of relativity, the standard Big Bang model has two possible destinies: continued expansion or recollapse. The nature of the final catastrophe (if catastrophe indeed awaits us) depends first on which comes to pass. If the density of matter and radiation exceeds the "critical density" of about 10^{-29} grams per cubic centimeter, the universe is "closed" and will eventually collapse; otherwise the universe is "open" and will continue to expand.

A count of the visible stars, galaxies, dust, and so forth suggests the actual density is only about ten percent of critical, so the universe should expand forever. But there must always be more mass than we actually see, since we cannot count everything, so the ten percent is if anything an underestimate. Furthermore, many theoreticians subscribe to the attractive inflationary model, in which the universe expanded shortly after birth to a state almost exactly on the borderline between open and closed. In this case the universe might well recollapse, but only after an essentially infinite amount of time. I will assume this variant is equivalent to continued expansion, and that if the universe is truly meant to recollapse, it will begin to do so in the uncomfortably short time of 15 or 20 billion years from today. As you see, our crystal ball on the future is sufficiently cloudy that we must consider each case separately.

In a short paper written in 1969, Martin Rees, a leading astrophysicist at Cambridge University, drew the first sketch of the End in a collapsing universe. Today we see a universe filled with stars, galaxies, clusters of galaxies, and even superclusters. Virgo, the nearest and most famous cluster, contains about 2,500 galaxies and lies at a distance of about 20 megaparsecs (65 million light-years). By cosmological standards this is quite close—the radius of the observable universe is approximately 5,000 megaparsecs. In general, clusters contain several hundred galaxies each and are separated by tens or hundreds of megaparsecs. They fill very roughly one percent of space.

Suppose the great collapse began. Once the radius of the universe had shrunk to one fifth its present radius (a process that would take nearly 14 billion years if the universe is today 15 billion years old), the volume would have decreased by a factor of 125. Galactic clusters would now fill 100 percent of space, and consequently would have merged into one. If there are a billion galaxies in the universe, any astronomers living at this time would see two colliding about every 100 years, with the frequency of collisions rising rapidly.

When the universe had contracted by another factor of ten in radius (a factor of fifty or so from the present size), the average density of the material in the universe would approach one particle per cubic centimeter—the density that exists today within the galaxy. At this stage the concept of galaxy would lose all meaning. Astronomers would not see the familiar spirals and ellipses that contemporary astronomers observe through their telescopes, but would merely record a universe of stars, more or less uniformly scattered across the heavens. They would probably not observe many giants and supergiants, since most of these would have burnt out long ago. Surviving would be white dwarfs and low-mass stars of the main sequence, which have long, stable lifetimes.

Of course, a few stars would still be forming—young bright stars that could give rise to life. But since galaxies have merged, stars would no longer be gravitationally bound to any original parent galaxy, as the sun is now bound to the Milky Way; instead, they would have detached themselves and begun to wander aimlessly throughout space like gas molecules in a teakettle. A civilization circling one of these stars might eventually find its sun destroyed by a near-encounter or collision with another wandering vagabond. You can envisage an astronomer watching in horror as a white dwarf sweeps through his solar system at a velocity near the speed of light, swings, by his sun, and disrupts it by tidal forces or causes it to explode by triggering new nuclear reactions. The astronomer will likely never make another observation.

Actually, this is the optimistic scenario of a science-fiction writer. Stars are tiny objects and will not begin to collide frequently until the very late stages of collapse—when the universe is only one hundred-millionth or one billionth of its present radius. By that time stars will almost certainly not exist. What has been left out of the preceding discussion is the cosmic microwave background radiation, the radiation left over from the Big Bang itself. As the universe expands, this radiation cools, and is presently observed at a temperature of 2.7 °K. Conversely, once the universe begins to collapse, the temperature of the microwave background will rise. The temperature is inversely proportional to the size of the universe, so when the cosmos has shrunk to only a thousandth of its present size, the microwave background will have reached 2700 °K, which is approaching the surface temperature of an average star. The sky will appear almost as bright as the sun. On the other hand, the situation will have grown very bleak.

A star must radiate from its surface the energy that is generated by the nuclear furnace at its core. This happens efficiently when the surroundings are at a very low temperature—as they are today. But if the microwave background approaches the surface temperature of the star, then to continue to radiate energy at the proper rate, the sun must get larger or increase its surface temperature (or both). Thus you can imagine a stage in the universe's collapse when stars begin to swell as the temperature of the background soars. This stage will not last long. Once the temperature of the microwave background exceeds that of the star, the star's surface layers will begin to boil off. This temperature is also close to that needed to ionize many of the elements that constitute the star. Electrons are stripped from the protons and neutrons they orbit, leaving a soup of atomic nuclei and free electrons.

Actually, things are worse. Any radiation, like light or microwaves, exerts a certain characteristic pressure. When the microwave background temperature exceeds the stellar temperature, the radiation pressure of the background will also exceed the surface pressure of the star. The star will be compressed, more and more as the background pressure rises. Eventually the star may implode, or, when the background temperature reaches nuclear ignition temperatures, the remaining stellar fuel could be detonated and the star disrupted in nova or supernova fashion.

A civilization might bury itself underground to avoid the increasingly severe microwave burn, but there is no escape. As the Big Crunch approaches, the temperature will climb indefinitely. Near the end they will remember Tom Lehrer's words to a terrestrial holocaust: "We will all fry together when we fry." And they will be right. No *deus ex machina* will be able to save those present from cosmic incineration. For perhaps one hundred thousand years after carbon-based life has been ionized, the collapse will continue. In the final instants before the Big Crunch, increasing numbers of exotic particles will be produced, just as new particles are discovered with each new and more powerful generation of earthbound accelerator. Photons will collide to produce electron-positron pairs, then neutrons and protons, eventually quarks, and finally particles that have yet to be discovered because their masses are far greater than anything that can be produced even by the proposed Superconducting Supercollider.

It could be that the very last stages of this scenario would be altered if the time before recollapse is so long that most stars and galaxies have been processed into black holes, a possibility I will discuss in more detail later. Then the black holes will begin to merge in the final seconds and swallow up more and more of whatever remains. The final instants will not be dominated by the proliferation of exotic particles, but by larger and larger black holes that finally engulf the entire universe at the instant of the Big Crunch.

Here, as Gogol might have said, events become befogged, and no one knows what takes place thereafter. The Big Crunch, according to Einstein's theory, is a singularity, a time when the universe's temperature, pressure, density, and all other physical quantities become infinite. Once a singularity is reached, the story ends. There is no way out. But many physicists feel that singularities are signs of defective theories, not natural artifacts, and that when we get close to the Big Crunch, Einstein's relativity should be repaired by wedding it to quantum mechanics. More precisely, at 10^{-43} seconds before the Crunch, quantum forces are expected to become as large as gravitational forces and must be taken into account. Unfortunately, a proper union of quantum mechanics and gravity continues to elude researchers, but it is hoped that quantum effects—which can cause negative pressures—may be able to halt the gravitational collapse of the universe and cause a "bounce." If this is true, then a new Big Bang cycle will begin.

The search for such an oscillating universe is not new. About fifty years ago, Richard Tolman of Caltech showed that the entropy during the universe's collapse phase should be slightly higher than that of the preceding expansion phase. For our purposes we can regard entropy as a measure of the waste heat produced by most physical processes, of energy that cannot be used (in contrast to the "free" energy available for consumption). In relativity theory a higher entropy implies a larger expansion (or contraction) rate. So if the universe bounced at the Big Crunch, the new Big Bang would start out with a higher entropy and higher expansion rate than the previous Big Bang. The universe would then expand to a greater maximum radius before recollapse than the universe that preceded it. Each cycle would produce a larger and larger universe. Rather than damping out the bounces, the increase in entropy actually amplifies them, a behavior that is exactly contrary to intuition. When you ride a bicycle you must keep pedaling; otherwise the friction between the bearings, which produces entropy, eventually dissipates the energy of motion and the bicycle grinds to a halt. In Tolman's oscillating universe, the production of entropy speeds things up.

Actually, Tolman's model makes two unjustified assumptions. The first is that somehow the singularity at the Big Crunch is avoided, and the second is that the entropy at the end of one cycle is equal to the entropy at the beginning of the next. One can indeed find models of the universe that do bounce (without resorting to quantum gravity), but these models are not very realistic in their details and tend to be ruled out by observational evidence. To date, all physically reasonable models end in a singularity. Consequently Tolman's phoenix may or may not rise from the ashes.

The question of entropy leads directly to the next scenario for the fate of the universe. Many physicists who ponder these matters feel claustrophobic in a recollapsing universe, and prefer the more optimistic alternative that the universe will continue to expand forever, or not recollapse for such a long time that it might as well be forever. But even such optimism is tempered by the second law of thermodynamics. The unavoidable and inexorable second law tells us that entropy always tends to a maximum in certain types of systems. As early as 1854, the German physicist Helmholtz realized the implications: the universe was using up all its free energy and would eventually run down like an unwinding clock. Stars would exhaust the available fuel, and finally the universe would sputter out to a lifeless cinder in a maximum-entropy state of constant temperature (at which point all evolution would necessarily cease).

The Heat Death of the universe, as this situation was called, was profoundly disturbing to both theologians and evolutionists, who believed that the universe must progress from a lower state to a higher. In his *Autobiography* Darwin wrote, "Believing as I do that man in the distant future will be a far more perfect creature than he now is, it is an intolerable thought that he and all other sentient beings are doomed to complete annihilation after such long-continued slow progress."

Darwin and others might have found solace in the fact that the second law does not apply to the universe as a whole. In a closed universe it cannot even be properly formulated; in an ever-expanding universe there is no maximum-entropy state and the temperature never reaches a constant value. So modern cosmologists who have reflected on the ultimate fate of an open universe have not been burdened with the assumption that its final resting place is a maximum-entropy graveyard. But I am not sure that you will derive any more comfort from the scenarios to follow. Most of what I am about to describe is based on the work of Jamal Islam, now of the University of Chittagong in Bangladesh, who in 1977 was the first to draw a picture of the distant future based on modern scientific knowledge, and Freeman Dyson, at the Institute for Advanced Study in Princeton, who further developed Islam's line of thought. Their apocalypse is somewhat protracted.

The rate at which a star burns depends on its mass; the lower the mass of the star, the slower it burns. Even the lowest-mass stars will have exhausted their nuclear fuel within the next 100 trillion (10^{14}) or so years. (As a standard of comparison, I will assume that the universe today has been expanding for roughly ten billion [10^{10}] years.) New stars are always forming, but the rate of star formation decreases as the raw materials—principally interstellar hydrogen—are used up. No one is certain how fast the construction material is being depleted, but some estimates give five billion years for half of the remaining supply to be used up. Five billion (5×10^9) is much less than 10^{14}, so even in an optimistic case we might expect all stars to have completed their nuclear burning within the next 10^{14} years. At that stage they will all have collapsed into white dwarfs, neutron stars, or black holes. By then the universe will have expanded to approximately 500 times its present size, and will have grown very dim.

Just as in the recollapsing universe scenario, there is always a chance that two stars will collide, or nearly collide. For example, a vagabond star may pass between the earth and the sun, and in the process detach the earth from its present orbit. Unlike the collapse situation, however, here the universe is expanding and stars moving ever farther apart; consequently we expect near encounters to be exceedingly rare. An easy calculation shows that the earth will almost certainly be detached from the sun within 10^{15} years. Since this holds for all stars, by the time the universe is 100,000 times as old and 2,000 times as large as it is now, most planetary orbits should be severely disrupted.

If two potentially colliding stars should approach one another to within a million kilometers or so (that is, so close that they almost touch), one may be flung around with such a high velocity that it actually escapes the galaxy. The time scale for this process to be effective turns out to be about 10^{19} years. In other words, when the universe is a billion times as old as it is now, close encounters will have "evaporated" most stars from the galaxy altogether. This evaporation process takes energy out of the galactic system, so the remaining stars in the central regions tend to fall together. The final result is likely to be a supermassive black hole of, say, 10^{10} solar masses, surrounded by a halo of dead stars.

This is not necessarily the end. Orbits also decay by the emission of gravitational radiation. Just as any accelerating charged body emits electromagnetic radiation, any accelerating massive body emits gravitational radiation. The earth, as it orbits the sun, is accelerating, and hence emitting gravitational radiation. The energy loss causes the orbit to decay, and given enough time, the earth will spiral into the sun. This time comes out to be about 10^{20} years. We see that it is 100,000 times longer than the time in which the earth is likely to be detached from the sun by an encounter with another star, so probably there is nothing to worry about. It is also ten times longer than the time in which we expect the sun to be evaporated from the galaxy, so before gravitational radiation can have an effect, the earth would probably be wandering around lifeless among dead stars and the sun's corpse would be orbiting in the galactic halo. But if it should just happen that the sun is ejected from the galaxy with the earth still in tow, then eventually the two would spiral together until they coalesced.

Events are just getting under way. The Islam-Dyson type of scenario assumes that matter is stable, which is not a currently fashionable belief. According to contemporary Grand Unified Theories (GUTs), protons and neutrons, which compose all ordinary matter, are unstable and should decay into electrons, positrons, neutrinos, and photons. These lighter particles are assumed to be stable and not decay further. In what is known as the minimal SU(5) GUT, the time scale for proton decay is about 10^{31} years. Unfortunately, experiments designed to detect proton decay should have discovered the phenomenon if SU(5) is correct, and to date they have failed. Should SU(5) be ruled out, there are two alternatives: either another GUT must be found that predicts a longer proton lifetime, or GUTs are in general incorrect and the proton is absolutely stable.

I will now depart from the Islam-Dyson route and assume that some GUT is correct. For illustrative purposes I will also assume that protons and neutrons (collectively termed baryons) will decay with a mean lifetime of 10^{31} years. Recall that after about 10^{12} years most stars will have ceased their nuclear burning and will have collapsed to white dwarfs, neutron stars, or black holes. In the absence of baryon decay a white dwarf will thereafter cool to a "black dwarf" at 1 °K within 10^{20} years. However, a simple calculation first performed by Gerald Feinberg at Columbia shows that the energy released by proton decay is enough to keep black dwarfs at approximately 5 °K for the remainder of the proton lifetime. Neutron stars will be heated to 100 °K, and the earth to a crisp .16 °K. An astronomer pausing to take stock of the situation after 10^{20} years would see perhaps 10% of the galactic mass concentrated in the supermassive black hole mentioned above, and 90% distributed among black dwarfs and neutron stars glowing at a few degrees Kelvin, with trace abundances of hydrogen and helium filling the interstices. This state of affairs will last until 10^{31} years, by which time 37% of the mass of all the stars and planets will have disappeared owing to insidious proton decay. At 10^{32} years, all but about .005% of the earth's mass will have disintegrated, and the planet will have shrunk to the size of Vesta, one of the larger asteroids. By 10^{34} years the earth will have dissolved into a sea of positrons and electrons. So will any life based on neutrons and protons.

On that day of wrath, the electrons and positrons that are the last remnants of earth and sun will annihilate each other to produce photons. Afterwards the universe will consist only of these stray glimmers, black holes, and a dilute electron-positron plasma, the decay product of the interstitial helium and hydrogen traces that had originally escaped incorporation into stars or black holes.

The black holes will eventually give their last hurrah. According to the celebrated result of Stephen Hawking at Cambridge University, black holes emit radiation like any hot body. In the process they lose mass, until they end their lives in a colossal burst of x rays and gamma rays. Under the Hawking process, a solar-mass black hole will evaporate completely in 10^{66} years, while a supermassive black hole of one tenth the mass of the galaxy will evaporate in 10^{96} years. If galactic-mass black holes coalesce into supercluster-mass black holes, these will evaporate after 10^{117} years.

The fate of the electron-positron plasma has been investigated by John Barrow of Sussex and Frank Tipler of Tulane, and by Don Page and Randall McKee at Penn State. If the universe is very open, it will be expanding fast enough that the electrons and positrons are always pulled away from each other and never meet to annihilate. On the other hand, in a "flat" universe, one just on the border between closed and open, the expansion rate gradually approaches zero. In this case, the electrical attraction between the electrons and positrons will finally become strong enough to overcome the effects of expansion. Electrons and positrons will then begin to orbit one another as atoms of positronium, an element created in earthbound laboratories decades ago. Page and McKee calculate that this positronium will form when the universe is about 10^{71} years old and 10^{40} times its present size. Unlike their earthly counterparts, which are about 10^{-8} centimeters in radius, the size of a cosmic positronium atom will be about 10^{12} megaparsecs, millions of times larger than the currently observable universe. The electrons and positrons will circle each other at roughly 10^{-4} centimeters per century, and in the process release electromagnetic radiation, or photons. This radiation loss will cause the electron-positron orbit to decay, until the two particles finally meet and annihilate each other. During their long spiral inward, which Page and McKee estimate lasts 10^{116} years, each positronium atom will emit about 10^{22} photons, far more than the billion or so photons that now exist for every atom.

Nevertheless, it appears that not all of the electrons and positrons converge to annihilate, and that a small fraction will always remain to form positronium. This tiny fraction, in fact, makes up the bulk of the matter in the universe. And so, finally, after 10^{117} years, the universe consists of a few electrons and positrons forming positronium, neutrinos left over from baryon decay, and photons from baryon decay, positronium annihilation, and black holes. For this, too, is written in the Book of Destiny.

In what I have described above, I have assumed that GUTs are correct and that baryons decay on a time scale of 10^{31} years. While most physicists expect GUTs to be confirmed sooner or later, because there is presently no experimental verification of these theories we should allow the possibility that baryons are forever. Then Dyson gives us a somewhat different picture of the remote future.

Stained-glass windows from medieval cathedrals, it is rumored, are thicker at the bottom than at the top. The rumor rings of plausibility, because on time scales of a millenium the chemical bonds in glass weaken, and a window will drip under the force of gravity. Whether a material acts like a solid or liquid depends merely on the time scales involved; any material will lose its rigidity if we wait long enough. Assuming that baryons are stable, Dyson calculates that iron will behave like a liquid after 10^{65} years. On longer time scales all objects—asteroids, planets, rocks—will become spherical under the action of their own gravity.

Similarly, all elements are radioactive. We do not normally think of nuclear fission taking place in gold, because quantum mechanics tells us that the probability is very small. But small probabilities become actualities, with enough patience. Dyson calculates that all matter will decay to iron, the most stable element, in a time of about 10^{1500} years. After this time, stars that had previously been black dwarfs will have been transformed into cold spheres of hard iron. When a further $10^{10^{76}}$ years elapses, iron stars will decay into neutron stars. During the process they will emit huge bursts of energy in neutrinos, and somewhat more modest bursts in x rays and visible light. As Dyson points out, the universe will still be producing occasional fireworks after $10^{10^{76}}$ years. I find this a comforting thought. However, it may be that everything decays into black holes in the much shorter time of $10^{10^{26}}$. These black holes will then evaporate in a negligible amount of time by Hawking radiation. After that there will be no more fireworks.

In this discussion of an ever-expanding universe I have left out one very important element: life. Life is the real reason physicists are concerned with the future, and the real reason they do not like to predict it. Traditionally, life plays no part in physics; life is messy and complicates things immeasurably. Physicists would prefer life not to exist. But it does exist, and one must try to incorporate it into any scenario of the future. Either life will remain cosmologically insignificant, as it is now, or the action of intelligence will finally alter the evolution of the universe in ways that are difficult if not impossible to predict. Both cases are interesting.

George Ellis of the University of Cape Town and I have recently thought a little about the pessimistic scenario, in which life remains cosmologically insignificant and evolves only until the stars burn out when the universe is one hundred to ten thousand times as old as it is now (see next chapter). If we imagine civilizations newly arising at this late stage, one thing is clear: it will be much more difficult to arrive at a reasonable picture of the universe than it is now. Galaxies will have grown far dimmer and more difficult to detect; the microwave background radiation (crucial for the establishment of a Big Bang theory) will have cooled to almost unmeasurable temperatures and gotten confused with the radio emissions from galactic dust; measurement of the primordial helium and deuterium abundances (that is, isotopes produced during the first three minutes and also crucial for the Big Bang picture) will be nearly impossible, because the helium produced later by stars will mask the original abundances.

In short, a science of cosmology may never develop; if it does, it is not obvious that cosmologists will ever arrive at a Big Bang theory of the universe. This raises an interesting question: are our theories true pictures of the physical world or merely reflections of the epoch in which we happen to develop them? (If you feel that present-day cosmology does not deserve to be termed science, then consider how lucky we are to be living now and not 100 billion years hence.)

Physicists like Dyson, being philosophical optimists, have considered the other extreme—the possibility that life can evolve and process information indefinitely. Humans require a warm environment in which to maintain their high metabolic rate and think. As the universe expands far past the lifetime of stars and cools toward absolute zero, life must necessarily slow its metabolism and move away from flesh and blood if it is to survive. The second law of thermodynamics tells us that to process information one must always dissipate some energy, so there is an absolute lower limit to the metabolic rate of an intelligent organism. And the disposal of metabolic waste heat becomes difficult when the creature's body temperature approaches that of its surroundings.

But Dyson proposes a way around this dilemma. He first defines a "subjective time" that depends on the metabolic rate; the faster the metabolism, the faster the subjective clock, and vice versa. If a creature hibernates for longer and longer periods of time, its biological clock is running intermittently, and it could dispose of its waste heat during its hibernations. And, since its subjective clock is running ever more slowly, it would feel as though it were living forever. During this infinite subjective time a creature could process an infinite amount of information, develop an infinitely large memory, and communicate with other creatures across arbitrarily large distances.

The question of energy consumption has been examined in more detail by Steven Frautschi of Caltech, who concludes that the best energy source for an expanding civilization would be black-hole radiation. (The giant positronium atoms of Page and McKee would be too dilute to collect, and the radiation from positronium decay would not supply enough energy for life to expand at the necessary rate.) We thus imagine an empire that farms black holes from an ever-increasing radius, tows them at substantial fractions of the speed of light to collection points, and traps the Hawking radiation in huge shells constructed for the purpose.

Unhappily, Frautschi discovered a fatal flaw in his own future: the tendency of solid structures to turn liquid over long periods of time would necessitate constant repair work, and the energy required for this would always exceed that reaped from the black holes. In any case, the likely decay of baryons makes the picture of a future civilization building black-hole shells highly improbable, at best.

Barrow and Tipler propose an alternative energy supply. In many cosmological models, the universe becomes increasingly irregular in the distant future. This irregularity (or anisotropy, as it is called) contains potential energy, which might supply a distant empire. Regardless of the final energy source, a more important question remains: what form will life take when the temperature approaches absolute zero and the last protons have decayed? It will not look like us. Perhaps intelligent creatures will consist of positronium atoms with radii larger than the currently observed universe. Bits of data could be stored by flipping the spins of the electrons or positrons. Whether it is in fact possible to organize positronium atoms with sufficient complexity to form sentient beings capable of reproduction is open to question.

You might wonder whether life can be prolonged indefinitely in a closed universe. Sadly, it appears impossible with the present laws of physics to change a closed universe into an open one, so that strategy is barred to us. Furthermore, Dyson originally concluded that an infinite subjective time was not possible in a closed universe; no matter how much a creature speeded up its metabolism to try to outrace the collapse, the Big Crunch always caught up with it in a finite amount of subjective time. More recently, Barrow and Tipler argue that in a very special type of closed universe this might not be true. They have formalized Dyson's basic idea into what they call the Final Anthropic Principle (see the previous chapter), which says that at the instant of this special Big Crunch, life will have arisen not only in this universe but in all possible universes, will have succeeded in processing an infinite amount of information, and will know everything that it is logically possible to know. A perversely optimistic apocalypse. And so, with an immortal smile at the Crunch of all Crunches, it's best to stop, for things have clearly gone out of sight.

CHAPTER 6

The Epoch of Observational Cosmology

By the time a science-fiction writer finally sits down at his desk to pen the opening lines of his great interstellar epic, he has often spent months, if not years, designing the future: Will the story take place on a planet? If so, how large will the planet be? What will be its composition? Its density? The strength of its gravity? Will life be based on carbon, silicon, or electromagnetic energy?

Any good science-fiction writer asks himself such questions, and if he is conscientious, the answers will not just be incidental to the story but considerably influence it. A planet with a much stronger gravity than earth's might give rise to short, stocky beings with powerful arms and legs. Perhaps on such a world the development of flying machines would be greatly delayed or never take place. If the planet, like Venus, were perpetually enshrouded by clouds, the ancients would walk with their heads to the ground and astrology would never develop—a blessing, to be sure, but neither would lunar calendars be invented or amateur astronomy.

One must be careful when designing planets or futures. On a cloud-covered world one could observe tides and consequently one could—with difficulty—deduce the presence of any moons and suns. Once technology had progressed far enough, radio and gamma-ray astronomy would become possible (because clouds are transparent at these frequencies), but any discoveries in these fields would probably be accidental, since, optical astronomy being nonexistent, no one would be scanning the skies to begin with. Thus one could eventually embark on a science of astronomy, but the picture of the universe you arrived at would differ considerably from our own.

Isaac Asimov once played this sort of game in his famous short story "Nightfall," in which the six suns in the sky of Lagash prevented its inhabitants from knowing that they lived within a "giant" star cluster. Once a millennium all six suns set, the truth is revealed, the Lagashans descend into madness, and civilization is destroyed. On the other hand, Larry Niven's *Ringworld* always seemed to me a failure in this regard (a minority opinion without a doubt), because few, if any, of the novel's adventures actually required the planet that Niven had developed.

In this essay I am going to use the science-fiction-writer's method to explore a question that might conceivably have some relevance to science: *at what stage in the universe's history could a science of cosmology develop*? By "science" I implicitly assume a discipline that has something to say about the real world. Astrology and scientology do not count. The question can be rephrased as: *what stage in the universe's history can be termed an epoch of observational cosmology*? Although the answer to this question should not cause descent into insanity, it should be provocative and hopefully bear on current theories of the universe.

EARLY TIMES

Suppose that you could observe the universe shortly after the Big Bang—shortly meaning from about 1 day to 100,000 years. What conclusions could you draw about your environment? Not many.

If you could build a thermometer, you would certainly be able to deduce that the universe was expanding, because the temperature would be dropping rapidly, along with the density and pressure. The universe's temperature is the temperature of the cosmic fireball radiation left over from the Big Bang itself—the same radiation that 10 billion years later is termed the cosmic microwave background radiation and is observed at 2.7 °K. One day after the Big Bang the temperature is about $3 \times 10^{7\circ}$ and halving roughly every four days. The fact that the temperature is so high means that this radiation would not be observed as microwaves (relatively cold photons) but as soft x rays (relatively hot photons).

Apart from this radiation, the only matter content of the universe would consist of some protons, deuterium, and helium—the light isotopes believed to have been formed three minutes after the Big Bang, when the temperature was high enough to support the nuclear reactions that could process neutrons and protons into heavier isotopes, and yet cold enough that the newly formed elements would not immediately break up again into the constituent protons and neutrons. As discussed in Chapter 4, primordial nucleosynthesis processes about 25 percent of the universe's mass into the common form of helium, helium-4; about 10^{-3} percent remains in deuterium, and similarly small, if not smaller, trace abundances are found in other light isotopes such as helium-3 and lithium. The other 76 percent is taken up by unprocessed hydrogen nuclei—protons.

The existence of the cosmic background radiation and primordial isotope abundances are in fact the main pieces of evidence that present-day cosmologists use in favor of the Big Bang, and without them such a model would be very difficult to substantiate. Yet it appears to be correct in its main features. The Cosmic Background Explorer (COBE) satellite verified in 1989 that the background radiation is of exactly the form predicted by the standard Big Bang model (a black-body spectrum). Today it is very hard to conceive of any explanation for the cosmic background radiation other than the Big Bang. Now, both the background radiation and isotope abundances would be even more apparent at very early times, so it might appear that any astronomer could easily establish a Big Bang theory.

However, observing anything would be very difficult. Galaxies, stars, and planets would of course not yet exist. Furthermore, at the high temperatures we have been discussing, the density of x-ray photons would be so great that they could not travel very far before colliding. At $3 \times 10^{7\circ}$, the so-called mean free path of a photon would be only 100 to 1,000 kilometers. Consequently, while an intelligent being might establish a Big Bang model, his concept of the universe would be rather limited—to a sphere roughly 1,000 kilometers in radius. Anything beyond this distance would be unobservable unless he developed neutrino astronomy. Neutrinos collide very rarely with themselves and with other particles. In fact, by the time the universe is one second old, neutrinos will have essentially stopped colliding with anything. So, if our astronomer could somehow collect neutrinos, he could gather information from distant parts of the universe. However, this is no easy task. Since neutrinos are so loath to interact with other matter, neutrino detectors are extraordinarily difficult to build. Indeed, neutrino astronomy exists only in the most primitive form today; detection of neutrinos from the supernova 1987A is its first major success.

At 100,000 years after the Big Bang, the situation becomes slightly more encouraging. Prior to this time the universe was so hot that the thermal energy of electrons exceeded the strength of the electrical attraction between electrons and protons. As a result, electrons were moving too fast to be bound with protons into atoms, and they remained separate—the primordial soup was highly "ionized." But when the temperature dropped below 10,000° the electron thermal energy became low enough that the electrons could drop into bound orbits around protons to form neutral atoms, a process that finished by about 4,000°. (For reasons unknown to anyone, this process is referred to as "recombination," a peculiar terminology as the electrons and protons were never combined to begin with.) It turns out that neutral atoms do not scatter photons as well as ionized atoms, and so the universe suddenly becomes transparent to light. There is still nothing to see. At 100,000 years after the Big Bang, the first perturbations in the primordial gas that will eventually become galaxies are just stirring. Though astronomy might be possible, it would not be very exciting.

This discussion makes the strong assumption that astronomers exist. It is of course very difficult to imagine any life form that could come into being one second after the Big Bang, when the temperature is tens of millions of degrees. In fact, it is probably safe to assert that life could not exist at the time. Consequently, no science of cosmology could develop for the simple reason that the universe would not permit the existence of cosmologists. Here is an example of *anthropic principle,* the subject of Chapter 4 of this book: the universe can only be observed at those stages permitting the existence of observers. Clearly, the anthropic principle makes the discussion of cosmology at very early times highly artificial.

The situation becomes marginally less dismal when the temperature drops to about 10 °K, or when the observable universe is roughly one billion years old and about five times as small as it is today. It is still unlikely that any sort of life could exist, because planets would not have formed, and probably not even galaxies or the earliest stars (known to astronomers as Population III stars). The oldest quasar yet found dates from about this period—say, for illustration, slightly later—and it would be surprising if many older ones are discovered.

So even if life could somehow exist, there would be very little to observe. The most important point is that an astronomer could not measure the redshifts of quasars or galaxies. As is well known, light from an object moving toward us becomes bluer; light from an object moving away from us becomes redder. The fact that most galaxies show redshifts is the third major piece of evidence that the universe is expanding. A universe one billion years old would be transparent to light, but without galaxies or quasars (or more than a few) the redshift clue to the Big Bang theory would be lacking. An astronomer could measure the temperature of the background radiation, but he could not directly measure, as he might have above, the rate at which its temperature was falling; by this time in the universe's history, the temperature changes appreciably only over time scales of billions of years. Since he would see only a constant temperature, there would be little reason to guess that the background radiation was left over from the Big Bang. In any case, at that time so much energy would be generated by various processes leading to galaxy formation that the background radiation would likely be masked.

The abundances of helium, deuterium, and the other light isotopes would perhaps be helpful. These isotopes, left over from the first three minutes, would still not have been processed into stars and galaxies, and would therefore be present in their original abundances. If a cosmologist could build a computer and program it to simulate primordial nucleosynthesis, he would confirm that the standard Big Bang model indeed produces the amounts of the isotopes he actually observes. The isotope abundances would then be the best clue pointing to the Big Bang.

Even so, it is hard to believe that anyone in such circumstances would come up with a Big Bang theory. One need only look to the early twentieth century to be convinced of this. By that time galactic redshifts *were* measurable, but there was still considerable debate as to whether the universe was expanding. (It was only in the 1920s that the Magellanic Clouds were unambiguously determined to be outside the Milky Way.) In the vastly more difficult circumstances of a one-billion-year-old universe, it seems quadratically doubtful that anyone could come up with a reasonable picture of the universe.

But like the discussion of the first day after the Big Bang, neither is this discussion of a somewhat older universe very convincing. Life could probably not exist only one billion years after the Big Bang. Even if it could, to observe the cosmic background radiation or the isotope abundances (which is done by spectroscopy) requires detectors. Traditionally, detectors (like telescopes or spectrometers) are built from heavy elements, which are made in stars. Because stars would not yet exist, neither would detectors. Cosmology would not progress very far unless someone could construct instruments solely of hydrogen, deuterium, helium, and perhaps a pinch of lithium.

LATE TIMES

We now, some ten billion years after the Big Bang, live in the epoch of observational cosmology. Above I argued, primarily on the basis of the anthropic principle, that this epoch probably does not extend back to an era when the observable universe was even a factor of five smaller than it is now. How far into the future can we expect the epoch of observational cosmology to extend? The future, as usual, turns out to be more interesting than the past. Because I assume, for the moment, that life will continue to exist into the indefinite future, the answer to this question will no longer require the anthropic principle.

Imagine the universe to be about 300 billion years old (about thirty times as old as it is now) and about ten times as large. In Chapter 5,

I mentioned that star formation is slowing because the raw materials—principally hydrogen—are being depleted as new stars are made; some estimates give only 5 billion years for half of the remaining material to be exhausted. Star formation may therefore not be taking place 300 billion years from now, or not at a high rate. Nevertheless some low-mass stars may continue to burn for another 100 trillion years. I will make the optimistic assumption that in 300 billion years galaxies will be roughly as bright as they are now.

As the universe expands, strangely enough, the redshift from a receding galaxy *decreases*. (This result is counterintuitive, but is basically a result of the fact that the universe's expansion rate is slowing down, so the redshifts decrease with time.) This means that the radiation from the observed galaxy actually becomes more intense. On the other hand, since at 300 billion years the galaxy is ten times as far away as it is now, its apparent size is ten times as small. The result is that the apparent brightness of a given galaxy is about 100 times less than it is now. Many stars will have ended their burning, and if star formation itself has ended, a galaxy's luminosity will be even lower, by more factors of hundreds or thousands. The upshot of this discussion is simple: galaxies that are presently easy to detect will become much more difficult to observe.

Because as time goes on light reaches us from ever greater distances, more and more galaxies become observable to us as the universe ages. But such galaxies are ever farther away and so are dimmer and dimmer. Thus, even though as the universe expands more and more objects can be observed, they become harder and harder to detect. It is easy to imagine a time in the distant future when galaxies would be so dim that an astronomer reaching our technological level could not conclude they are external to the Milky Way. In fact, he might term the faint patches nebulae, as nineteenth-century terrestrial astronomers did before they realized the smudges were other galaxies like our own.

With distant galaxies so dim, their spectra would be difficult to obtain, and it might become impossible to determine galactic redshifts. An astronomer would then lack the key piece of evidence that the universe is expanding. Quasars, thought to be violent, active galaxies in the early stages of their evolution, would presumably have vanished altogether, eliminating that piece of evidence that there *was* an early stage to the universe.

At some stage, perhaps when the universe has expanded by another factor of ten, all galaxies that are not gravitationally bound to the Milky Way will have receded to the point where they have disappeared from view. Only those galaxies gravitationally bound to ours and therefore not receding (that is, those galaxies in our immediate cluster) would remain. An astronomer might then conclude he lived in a static island-universe, consisting of a relatively small number of galaxies. The history of cosmology, in which some of the earliest cosmological models *were* static island-universes, shows that this is a very natural assumption to make.

What about the cosmic microwave background? Could an astronomer 300 billion years from now measure it? First, the temperature of the background radiation would have fallen to .27° from its current 2.7°. The lower temperature is about the error in the 1989 satellite measurements, so an accidental discovery with the primitive equipment of 1965 (when Penzias and Wilson stumbled onto the radiation) would seem unlikely. Furthermore, the factor of ten decrease in the temperature means that the frequency at which the radiation is detected (now in the microwave region) would have also fallen by a factor of ten, and that the intensity of the radiation would have fallen by a factor of 1,000. The thousandfold decrease in intensity implies that in order to detect the background, a receiver would have to be 1,000 times more sensitive than those used today. In any case, when the background's peak frequency dropped by a factor of ten, it would be radiating at the same frequency as galactic dust, so there would be almost no reason to suspect that the signal came from the Big Bang at all.

Given that galactic redshift data would at best be poor, perhaps nonexistent, and that the microwave background might never be discovered, it seems plausible that a cosmologist living 300 billion years from now might never posit a Big Bang theory. Rather, as already mentioned, he might conclude he lived in a static island-universe.

Could the isotope abundances save him? Probably not. By this time all the original abundances would be wiped out. That is, the helium, deuterium, and lithium produced by the Big Bang and measured today would have all been incorporated into stars and reprocessed into heavier elements. Then, as stars became supernovae, the new mix would have been scattered into space. This recycling would happen not just once in 300 billion years, but probably a dozen times. If the future cosmologist computed, as we do, the abundances resulting from the standard Big Bang model, he would calculate that about 10^{-5} of the mass of the universe should be concentrated in deuterium. But observations would refute him—owing to stellar recycling he would find no deuterium whatsoever.

How would a cosmologist account for the isotope abundances that were measured? The history of cosmology, before the Hot Big Bang model won out over its competitors, provides many alternative explanations. Especially popular some years ago were the so-called Cold Big Bang models, in which the universe started out as pure hydrogen at a temperature of absolute zero. Gradually the hydrogen condensed into stars, which subsequently underwent nuclear ignition. In these models the observed element abundances are the sole product of stellar nucleosynthesis. Traditionally, the problem with the Cold Big Bang models was predicting the correct element abundances without elaborate juggling of parameters (the Hot Big Bang accomplishes this fairly naturally). But in the far future, when the primordial abundances will have been masked by stellar reprocessing, a cosmologist's only choice might be to invoke stellar astrophysics to account for the observations.

At late times, it would be difficult to explain not only galactic redshifts, the cosmic background radiation, and the isotope abundances, but also the entire thermodynamics of the universe. For instance, it is frequently suggested that the second law of thermodynamics is a result of the Big Bang. The second law, discussed extensively in Chapter 3, asserts that entropy of an isolated system must never decrease. Indeed, the overall entropy of the universe seems to be continually increasing. No generally accepted explanation for the second law exists; however, many cosmologists believe that if the Big Bang was a "zero-entropy" event, entropy—so to speak—had nowhere to go but up. Of course, if in the far future it is impossible to establish a Big Bang theory, such suggestions would never arise. The universe, now viewed as an uninterrupted cooling-off after the Big Bang, might be viewed as having more the nature of a candle suddenly flaring up from an unknown creation event. Proposals such as the one advanced by Ilya Prigogine (see Chapter 3), that the second law arises from microscopic considerations, might be viewed with more sympathy.

Throughout the discussion of the late universe, in order to avoid the complications of the anthropic principle I have assumed that life continues to exist. How reasonable an assumption is this? If, of course, civilization destroys itself through nuclear warfare or environmental neglect, then our epoch of observational cosmology will end abruptly. Otherwise, it is conceivable that our species can prolong its existence through various colonization strategies or the more extreme measures of Dyson discussed in the previous chapter. If the records of present and future cosmological achievement survive, our ancestors could refer to them as evidence for a Big Bang theory, but direct observational confirmation would be extremely difficult, and so the archival theories could not be proved or disproved. In this situation, cosmology might take on the status of a religion (which many would argue is already the case). There is also the possibility that new civilizations might arise on young stars and have to make observations for the first time. If the above arguments have any validity, they would not get very far.

VERY LATE TIMES

Of course the universe will not abruptly end when it has expanded a factor of ten or one hundred from its present size. There are essentially two possibilities, those discussed in Chapter 5: if the matter density of the universe is at or below the "critical value," the universe is "open" and will expand forever; if the density is above the critical value, the universe is "closed" and will eventually recollapse. In the latter case, depending on the exact value of the density, recollapse may take place after a relatively short amount of time, say another 20 billion years, or it may take place after a relatively long period of time, hundreds of billions of years or ten-to-the-hundreds-of-billions-of-years. No one knows.

If the universe recollapses after only 20 billion years, the epoch of observational cosmology will extend until that moment when the background radiation has heated up sufficiently to fry existing cosmologists. If the universe recollapses after hundreds of billions of years (or more), galaxies will recede from view for a long period, during which cosmology will go into hibernation; eventually galactic remnants will come back into view as the universe recollapses, and cosmology will experience a renaissance. (Here I assume that the recollapse takes place in less than 100 trillion years, so some stars may yet be burning.)

Galaxies containing surviving stars would naturally exhibit blueshifts (due to the universe's collapse) rather than redshifts. If they were bright enough to observe, the blueshifts might lead any newly emergent civilization to arrive at a Big Crunch model of the universe. But would there be any reason to suspect a previous expansion phase? If cosmologists could determine that the density of the universe was above critical (a difficult measurement), they might conclude that an expansion phase was possible, but there would be no direct observational evidence to indicate that one had actually taken place. The background radiation would be heating up, not cooling down, and the isotope abundances would long ago have ceased to bear any resemblance to their original values. Thus any civilization existing at that time without archival information could arrive at a totally incorrect picture of the universe. Civilizations with archival information would have the correct picture, but there would be no experimental reason to believe it.

At the other extreme, the universe could expand forever, or recollapse, but after such a long period of time that it might as well be considered infinite expansion. Such a scenario is possible in the popular inflationary universe model (see Chapter 4). According to inflation, the density is so nearly critical that if the universe does recollapse it will be only after 10-to-the-*n* ages of the universe, where *n* is a large number. In this case, as I have already discussed, all galaxies except those gravitationally bound to the Milky Way will eventually fade from view, and the microwave background will become undetectable. The universe model most likely to be developed by a cosmologist would be an island-universe cosmology.

Much of what I have said is highly speculative. If the ruminations have any correspondence to reality, then one would conclude that in our universe an epoch of observational cosmology extends from the time when the universe was about a factor of five smaller than it is now until a time when it is a factor of ten or so larger than it is now. In a closed universe there will be an additional period of observation during the collapse phase until the microwave background gets hot enough to disrupt life. Although the anthropic principle requires that the epoch of observational cosmology be a subset of the period during which life exists, it does not require that it exactly coincide with the period during which life would be found; apparently, the epoch of observational cosmology could be much shorter.

Throughout the argument I have made a tacit assumption: I have assumed that the universe is more or less described by the standard model and that we live in the epoch of observational cosmology. In the standard model (see next chapter) all parts of the universe look the same. But suppose on some very large scale (beyond what we can presently observe) the universe is not so uniform, and different regions have very different properties. The density may vary from region to region, galaxies may cluster in a different manner, and so on. As the universe expands, these new regions will come into view, and it will become evident that the standard model—which postulates perfect uniformity—was, after all, incorrect. In that case, our descendants would have to conclude that we did not live in the epoch of observational cosmology.

This raises the interesting proposition, to what extent are the questions we raise as cosmologists functions of our historical epoch? The inflationary universe provides a good example of the problem. Inflation was developed in part to explain why the universe is so flat—in other words, why the density is so nearly critical. But this state of affairs is clearly dependent on the epoch of observation. I mentioned in Chapter 4, that if one waits long enough, the equations of relativity will always change a nearly flat universe to one that is "curvature dominated," which means that the density will always be pushed away from the critical value. Thus the question inflation attempts to answer—"Why is the universe so flat?"—is a temporary dilemma and will eventually disappear.

It is a good lesson to bear in mind. If our picture of the universe is strongly influenced by the epoch of observation, then it is conceivable that universes could exist in which life arises only at times when observations lead to a deceptive understanding of cosmology.

ACKNOWLEDGMENT

"The Epoch of Observational Cosmology" is closely based on an article of the same name, which was coauthored with G.F.R. Ellis and appeared in *The Observatory* 107 (February 1987): 24–29. I thank George for his insights into this (and everything).

CHAPTER 7

Alternative Cosmologies

(WITH G.F.R. ELLIS AND RICHARD MATZNER)

ANY scientist knows that it is a dangerous game to make statements based on a single data point. A statistician would hesitate to predict the most probable outcome of a die toss until he or she had spent a sufficient number of hours at the gaming table. Physicists think they understand the behavior of hydrogen atoms because they have many identical atoms at their disposal and can study them under a variety of experimental conditions. Psychologists may have difficulty understanding the behavior of Ellis, Matzner, or Rothman individually, but observations of thousands of natural philosophers will reveal certain common properties. Geologists are in an admittedly uncomfortable position: they are stuck with one earth. But to some extent they can get information from other astronomical bodies, which, apart from any practical value, gives them confidence that statements made about the earth have a more general applicability. They are also able to place the earth's formation into the larger context of the origin of the solar system and are able to assume that the known laws of nature, such as conservation of energy and momentum, are applicable to their investigations.

Of all the sciences, cosmology—the study of the structure and evolution of the entire universe—occupies a peculiar position. The peculiarity stems from one basic fact: the universe is unique. Cosmologists are unable to vary any experimental parameters. Questions like "Is the total momentum of the universe conserved?" have no clear answer (and no clear meaning). Cosmologists cannot place the origin of the universe into a larger context. And unlike the statistician at a crap game they cannot claim that the universe is behaving in a probable or improbable manner; there is no practical way to roll another universe.

To circumvent such technical difficulties cosmologists create model universes that are meant to duplicate features observed in the real cosmos. To date, hundreds of cosmological models have been developed. For about twenty-five years the "standard" Big Bang model has been considered the most successful in accounting for the observations. Nevertheless, it does not account for all the observations, and so cosmologists have been forced to search for alternatives. What are the alternatives, and what considerations guide cosmologists in their search for them? As will become apparent, the uniqueness of the universe forces cosmologists to choose a favorite model as much by philosophical inclination as by data.

One powerful philosophical principle that has proved its worth in the history of science is Ockham's razor: the simplest theories are the best. In the standard model Ockham's razor finds expression as the Cosmological Principle: the universe is homogeneous and isotropic on all scales. The term homogeneous refers to any system that is translationally invariant, that is, it appears the same at all points. The term isotropic refers to any system that is rotationally invariant, that is, it appears the same in any direction around a fixed point. Loosely speaking, the Cosmological Principle states that the universe should be absolutely uniform everywhere. If it has a preferred direction, which would be the case if all the galaxies were moving in parallel to each other, it is anisotropic; if it is lumpy and full of irregularities, it is inhomogeneous (see Figures 7.1 and 7.2). It is important to note that isotropy about every point implies homogeneity, and so a completely isotropic universe must be homogeneous, but a homogeneous universe need not be isotropic.

The Cosmological Principle does not indicate whether the universe should be expanding, contracting, or static, but since about 1965 the term "standard model" has referred to a cosmology that started expanding at a hot Big Bang and whose subsequent evolution has been governed by Einstein's equations of general relativity. The standard cosmology is also known as the FLRW model, after Aleksandr Friedmann of the Soviet Union and Georges Lemaître of Belgium, who independently formulated the idea in the 1920s, and Howard Robertson of the United States and Arthur Walker of England, who extensively developed it in the 1930s.

As is well known, the FLRW model can have negative spatial curvature (like a saddle), positive spatial curvature (like a sphere), or neither. In the first case the model will expand forever and is therefore called open. In the second case it will eventually recollapse and is termed closed. In the third it lies on the borderline between the first two possibilities and is called flat. Although popular accounts frequently speak of the universe itself expanding forever or recollapsing, it is important to realize that this is a property of the standard *model*; whether the real universe behaves this way is a separate issue.

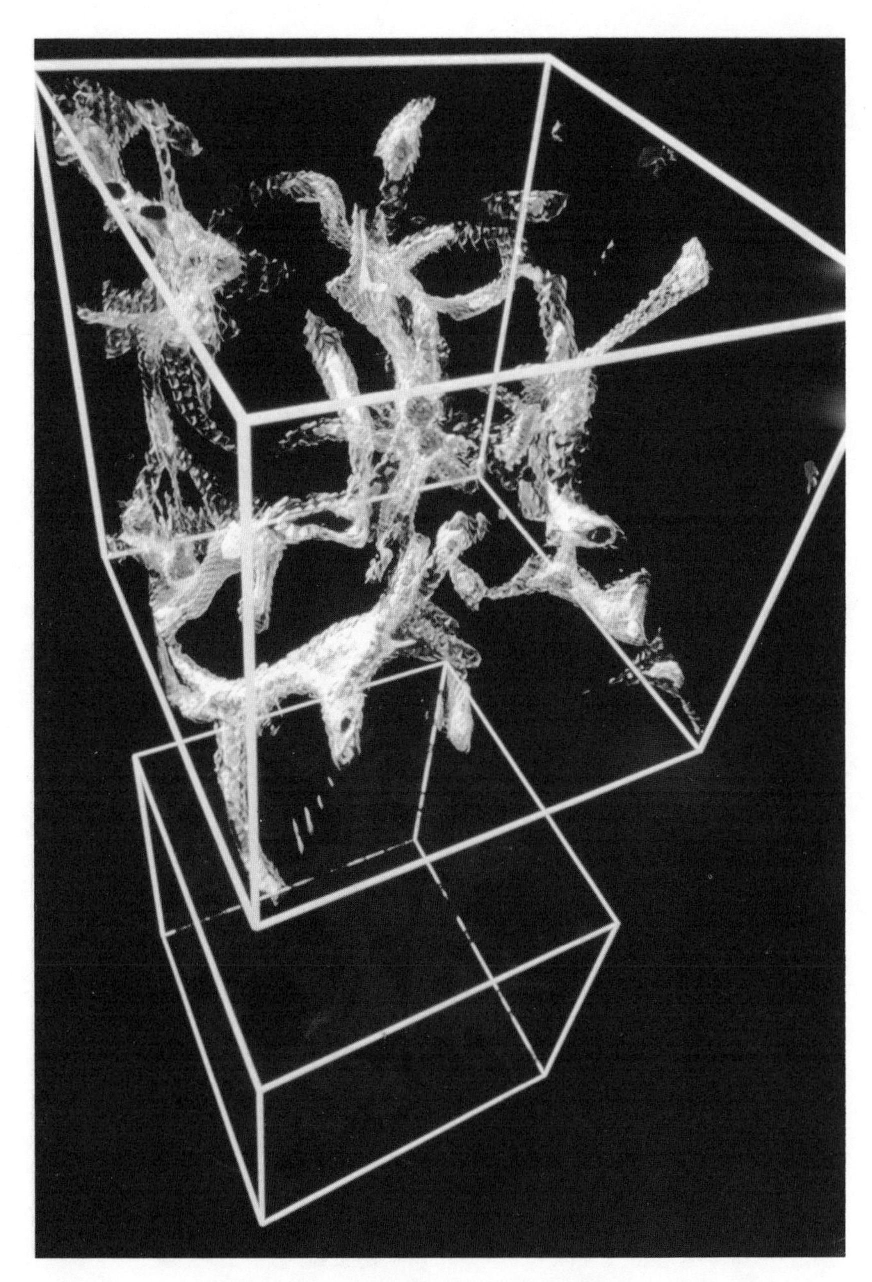

Figure 7.1 Joan Centrella's computer simulation of structure formation in the early universe. In this simulation filaments, sheets, and voids arise naturally. Similar structures are observed in the real universe, showing that the universe is, at least to some extent, inhomogeneous, or lumpy. The structures in Centrella's simulation are produced by placing substantial initial fluctuations in an isotropic nd homogeneous (uniform) background (see Figure 7.2). Perturbations of this size may be inconsistent with those arising from the standard cosmological model and suggest the need to investigate alternative scenarios.

Figure 7.2 Christmas wrapping paper demonstrates the concepts of homogeneity and isotropy. A perfectly homogeneous universe would be one that is *exactly* the same at every point—for instance, a perfectly uniform gas (not shown). Such a universe would be also isotropic, because from every point the universe would appear the same in all directions. If one considers just the blossoms on an infinite piece of wrapping paper (a), this universe is very nearly homogeneous, because all the blossoms are identical. It is also isotropic, because standing on any blossom one sees the same number of blossoms in all directions, of the same color and evenly spaced. Hence there is no preferred direction, the condition for isotropy. However, if one adds stems to the blossoms, "up" becomes distinguished from "down," and the universe is then *an*isotropic. It remains homogeneous, because all the flowers are still identical. (Isotropy always implies homogeneity—one can easily show this—but the reverse is not true. See "The Garden of Cosmological Delights" in *Science à la Mode*).

In wrapping paper (**b**), on the other hand, the different blocks have various colors and designs. This universe is therefore inhomogeneous and anisotropic. Nevertheless, if one takes units of 16 blocks (or 4, ignoring the color difference) the pattern is repeated. We can then say that on the scale of 16 blocks this universe is homogeneous. It remains anisotropic because the serrated edges in some of the blocks define a direction.

Although the mathematical description of the FLRW model was developed in the 1920s and 1930s, it was only after World War II that the physics of a hot Big Bang received extensive investigation. At that time researchers made two important predictions based on the model. The first is that radiation from the Big Bang should be observable today. In 1965 the discovery of the cosmic microwave back-

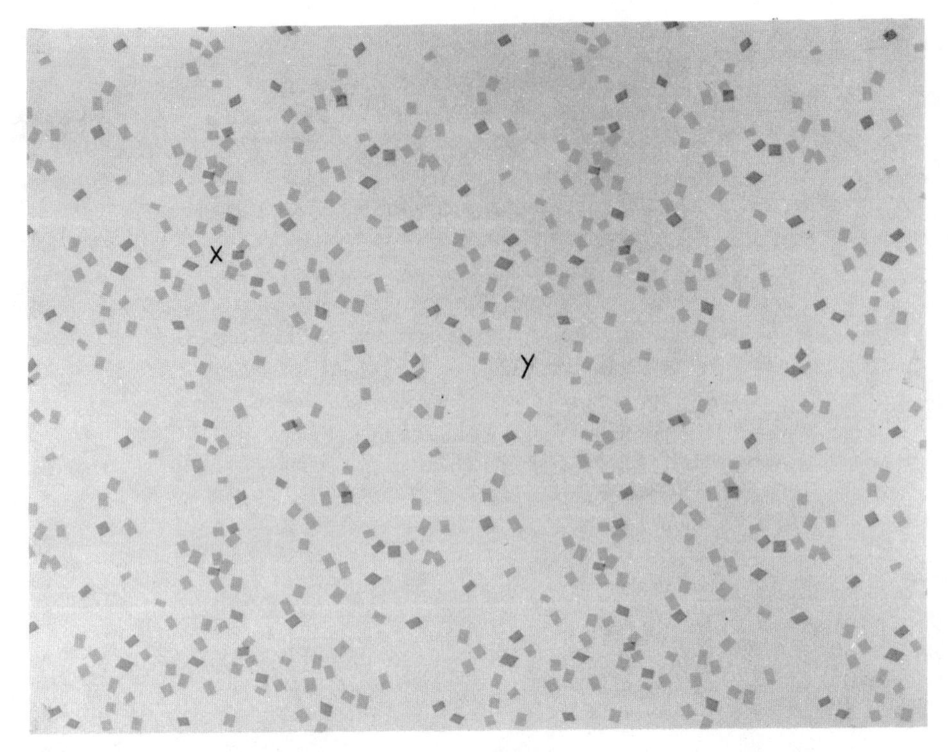

If you were located at point X on wrapping paper (c), you might conclude that the universe was approximately homogeneous and isotropic, since the blocks are scattered almost uniformly in the vicinity of X and there are roughly the same number of blocks in each direction. You would certainly not arrive at this conclusion if you were standing at point Y. Still, on the largest scales the pattern repeats itself, and so the wrapping paper is approximately homogeneous and isotropic at those scales. The standard cosmological model is *exactly* homogeneous and isotropic on *all* scales. The wrapping-paper designs show that exact homogeneity and isotropy are drastic simplifications. Can one assume the real universe is so simple?

ground radiation at about 2.7 degrees Kelvin verified this prediction. Repeated measurements (especially with the new Cosmic Background Explorer satellite, COBE) have shown the microwave background to be isotropic to better than one part in ten thousand, apparently justifying the Cosmological Principle.

The second prediction, refined by nucleosynthesis calculations carried out in the 1960s, concerns the composition of the universe. During the first three minutes after the Big Bang, the standard model processes about 25 percent of the universe's mass into helium, about

5×10^{-3} percent into deuterium, and comparable amounts into the other light isotopes. Such abundances have been observed in the real universe (see "Particle Accelerators Test Cosmology," by David Schramm and Gary Steigman, *Scientific American*, June 1988), with the result that many investigators refuse to take any alternatives seriously.

On the other hand, the Cosmological Principle is difficult to justify. In spite of the microwave background, observations clearly show that the matter distribution in the universe is not homogeneous and isotropic: galaxies and galactic clusters represent cosmological irregularities. Nevertheless, this objection has traditionally been swept aside with the reply that the universe is uniform on scales large enough that local wrinkles are negligible. Redshift surveys conducted over the past decade, however, have begun to make it more difficult to ignore the criticism. The surveys sample light from galaxies out to about 5 percent of the horizon distance (the distance light has traveled since the Big Bang) and show clearly that galaxies are not spread with exact uniformity but are arranged in irregular structures resembling sheets, filaments, and voids (Figure 8.1). In other words, on the largest scales that have been measured in detail the distribution of galaxies is inhomogeneous and anisotropic.

Faced with this dilemma, an investigator can either assume the universe began isotropically and gradually evolved structure, or began "chaotically"—that is, anisotropically and inhomogeneously—and evolved to a more uniform state. Disciples of Ockham's razor and the Cosmological Principle tend to take the former approach, arguing that the addition of anisotropy or inhomogeneity makes the models more complicated and philosophically unattractive. Coupled with the more immediate consideration that simple calculations cause fewer headaches than complex calculations, this school has perhaps been dominant in cosmology.

For a number of years, however, workers have realized that there are several serious difficulties with the "isotropic" approach, of which we mention three. The first is galaxy formation. If one assumes that galaxies are the end product of random density fluctuations in an otherwise uniform primordial soup that gradually grew under gravitational attraction, then in order to produce galaxies by the present epoch, the size of the density fluctuations must have been about one part in a thousand, 100,000 years after the Big Bang. Simple calculations suggest that such density fluctations would produce anisotropies in the microwave background of about the same size, that is, one part in a thousand. But, as mentioned above, measurements of the microwave background show that it is isotropic to one part in ten thousand.

A related problem concerns large-scale motions of galaxies. Observations carried out most recently by A. Dressler, M. Aaronson, and their colleagues out to about 3 percent of the horizon distance show that galaxies are not moving randomly but are streaming at a high velocity in a preferred direction. In an FLRW model one can calculate the velocities that would result from random initial fluctuations; the observed streaming motion is about twenty times too large.

The third problem was first clarified in 1956 by Wolfgang Rindler, now of The University of Texas at Dallas, and has become known as the horizon problem. Since no signal can propagate faster than the speed of light and the horizon distance is defined as the distance light has traveled since the Big Bang, a given particle cannot interact with any other particle separated from it by more than the horizon distance. One cannot see farther than one's own horizon. As the FLRW model is run backward toward the Big Bang, it turns out that the horizon distance shrinks faster than the average separation between particles, and so at early times any particle could interact only with a few neighbors. At the instant of the Big Bang itself, although the distance between particles went to zero, all particles were nevertheless causally disconnected: no particle was able to interact with any other.

In the past ten years this property has figured prominently in lists of the standard model's defects for the following reason: If different regions of space were causally disconnected at the Big Bang, then why did those various regions not end up with radically different properties? How did the universe "know" to start off isotropically everywhere? Clearly the horizon problem is somewhat different in nature from the previous issues. If one assumes the universe began as an FLRW model, then it is isotropic by definition and there is no problem. The tacit assumption being made by investigators is that the universe did not start off isotropically. In that case, it is difficult to understand by what mechanism the universe could have become more uniform, because the existence of horizons at the Big Bang prevented any interaction between particles.

Given that most cosmologists consider the horizon problem a real difficulty, the implication is that a study of nonstandard cosmologies is necessary to understand the possible origins of the universe. This school of thought, which traces its origins to Charles Misner of the University of Maryland, holds that it is not very reasonable to assume that the universe started in an exactly uniform fashion. If one imagines all possible initial configurations of the universe, those necessary for exact isotropy and homogeneity are very special indeed, and, it might be argued, infinitely improbable. We shall discuss this point more fully below, but here the problem of the universe's uniqueness is encountered in full force. Terms such as "plausible," "likely," and "probable" lose all meaning when applied to the origin of the universe.

Nevertheless, if one chooses to investigate chaotic cosmologies, the easiest approach is to allow the model to be anisotropic but homogeneous. For example, one might assume that the universe is expanding faster in one direction than another. The different expansion rates give a preferred direction and make the expansion anisotropic. The simplest model that does this is called a Bianchi I cosmology, after the nineteenth-century Italian mathematician Luigi Bianchi, who showed that there are nine ways that space can be homogeneous but anisotropic. Some of the Bianchi cosmologies can be regarded as anisotropic generalizations of the FLRW model. In the limit of exact isotropy Bianchi I becomes the flat FLRW model, Bianchi V becomes an open FLRW model, and Bianchi IX becomes the closed FLRW model.

Of all the Bianchi classes, Bianchi IX is the most complicated—not only do the three spatial axes expand at different rates but, unlike the cases of Bianchi I and V, space itself is anisotropic. (In I and V, only the expansion *rates* are anisotropic, not space itself.) Moreover, which axis is expanding fastest changes with time, and near the Big Bang at least one axis is always contracting. In addition, the spatial anisotropy varies with time. The behavior of Bianchi IX is more than unusually complicated: at early times it is mathematically chaotic. Nevertheless, a useful mental picture of Bianchi IX is to regard it as a cosmic washing machine that first churns in one direction, then another, infinitely many times back to the Big Bang.

Misner had hoped that the churning action would bring particles from all regions of space into causal contact with each other. In this way the horizon problem was to be solved. Viscosity, or frictional effects, between the streaming particles would then act to smooth out any initial anisotropies. Appropriately Misner termed the Bianchi IX model the "Mixmaster" universe, but subsequent studies showed that the Mixmaster action was not effective; horizons still existed in most directions and particles were not brought into causal contact with one another. Therefore the Mixmaster universe failed to solve the horizon problem and make it possible for physical mechanisms to smooth out initial irregularities.

Anisotropic cosmologies have also encountered difficulties when confronted with observations of the abundances of light nuclear isotopes. Computations based on the standard model predict that about 25 percent of the mass of the universe should be concentrated in helium, in good acccord with observations. Calculations in Bianchi I models over the past twenty years have shown that anisotropy tends to increase the helium abundance above the value produced in the isotropic case; even small amounts of anisotropy push helium above observational limits. Cosmologists have used the Bianchi I results to rule out substantial anisotropy at about one second after the Big Bang when the processes leading to helium formation began. It should be pointed out, however, that it is far from obvious that the more complex Bianchi models also increase helium. Recently we have shown, with certain caveats, that Bianchi types V and IX do indeed raise helium, but claims made on the basis of Bianchi I alone cannot be accepted as generally valid.

Although the Bianchi models are much more complicated than the FLRW model, they are still highly simplified in that they assume that space is homogeneous—the same everywhere. The next level of complication is to investigate inhomogeneous models. As one might guess, however, the equations governing the behavior of inhomogeneous cosmologies are extremely difficult to solve. In order to get exact mathematical solutions, drastic simplifications must be made. For example, one can consider an expanding isotropic universe that is riddled with spherically symmetric holes. Such a cosmology, aptly termed the Swiss-cheese model, was introduced by Einstein and Strauss in an attempt to understand whether the expansion of the universe caused the solar system to expand as well. The assumption of background isotropy with spherically symmetric holes makes the problem simple enough that exact solutions can be obtained; nevertheless the Swiss-cheese model is inhomogeneous because some regions of spacetime have holes and others do not.

Still, such a model is not considered by most cosmologists to be very realistic. Investigations of "realistic" inhomogeneous models have usually been carried out numerically with the help of computers or, before the widespread use of computers, in extremely crude fashion. In the crucial area of primordial nucleosynthesis, early results differed somewhat, but inhomogeneous models were generally thought to raise the primordial helium and deuterium abundances above observational limits, as did anisotropic cosmologies. Recently there has been a considerable debate on this claim, a debate to which we shall return below.

Because early investigations into anisotropic cosmologies appeared to give results at odds with observations and serious calculations involving inhomogeneous models are so difficult, after the late 1960s the study of nonstandard models languished in obscurity. In the last few years there has been a resurrection of interest in alternative models for one reason: inflation.

In 1981 Alan Guth, now at the Massachusetts Institute of Technology, proposed his famous inflationary scenario to solve, in large part, the horizon problem (see "The Inflationary Universe," by Alan Guth and Paul Steinhardt, *Scientific American*, May 1984). Recall that at early times in the FLRW model, the horizon was smaller than the interparticle separation, with the result that particles could not interact with each other to smooth out any potential nonuniformities. In Guth's model, at about 10^{-35} seconds after the Big Bang the universe undergoes a brief but enormous period of expansion ("inflation"), which arranges things so that the interparticle separation becomes less than the horizon size. This a necessary condition for the removal of any irregularities: particles can interact, although Guth's model does not specify the nature of the interaction. Still, inflation allows the universe to evolve from an initial chaotic state to the present nearly isotropic state.

Interestingly enough, Guth carried out his original computations in the FLRW model. But as we said above, the FLRW model is by definition isotropic, and so there is no horizon problem. The most that can be said for Guth's argument is that it showed inflation is compatible with isotropy. Guth mentioned this problem in his paper and gave a plausibility argument to show that exact initial isotropy was not required. Nevertheless, the actual calculation, and those carried out for several years thereafter, were all done in the standard model.

Gradually researchers have begun to realize that in order to prove that inflation really does provide the necessary conditions to homogenize and isotropize the universe, one must begin with anisotropic and inhomogeneous initial conditions. A number of such calculations have been made, and the results have been mixed. In an important paper, Robert Wald of the University of Chicago showed that once inflation gets under way in the Bianchi models, it will eventually isotropize the universe. Two of us (Ellis and Rothman) showed that inflation indeed takes place in all the Bianchi models (with certain caveats) and that the isotropization strictly speaking takes place before inflation begins. Lars G. Jensen and Jaime A. Stein-Schaubes, then of Fermilab, have performed an analysis similar to Wald's on certain inhomogeneous cosmologies and find that once inflation gets under way it performs as desired.

On the other hand, in an inhomogeneous numerical model developed by Hannu Kurki-Suonio, Joan M. Centrella, James R. Wilson, and one of us (Matzner) inflation takes place for certain initial conditions and not for others. A universe investigated by A. K. Raychaudhuri and Bijan Modak of Presidency College at Calcutta inflates, but the anisotropy is not damped out. Yet other models encounter singularities—regions where the physical quantities under investigation become infinite—and the universe literally goes up in smoke.

The diversity of results has led to the following essential question: Is the "measure" of initial conditions that leads to inflation large or small? In other words, is the number of initial configurations that undergo inflation large or small compared with the total number of conceivable initial configurations? Proinflationists tend to argue that the measure is large—"most" models undergo inflation. What, if anything, does this mean?

It turns out that 15 parameters (governing, for instance, the curvature of space and the anisotropy) are all that is necessary to describe the initial configuration of any homogeneous model, including the Bianchi models and the FLRW universe. If one plots each parameter along an axis, one can then imagine a 15-dimensional space of all possible initial conditions for homogeneous models. Since each parameter can take on an infinite number of values, there are an infinite number of possible initial conditions in this space. A particular initial condition is chosen by fixing the values of the 15 parameters and represents one point in the space. The FLRW model is represented by the plane formed by the two axes on which all the parameters governing anisotropy and curvature are set to zero. A two-dimensional plane cut from a 15-dimensional space represents a vanishingly small subset; the total number of available points is 13 factors of infinity larger. For this reason we stated above that the odds of the universe starting out with perfect isotropy are infinitely small.

But homogeneous models themselves form only a small subset of all possible worlds. To describe a fully inhomogeneous model at a given time requires 15 parameters at each of the infinite number of points in space. Therefore to describe the initial conditions of all fully inhomogeneous cosmologies would require 15 infinities of parameters, each of which could assume an infinite number of values. The probability of an isotropic Big Bang is very small indeed. Again we are confronted with the uniqueness of the universe. It may well be true that it is meaningless to speak of the odds in creating the universe. On the other hand, by any measure we are able to conceive of, such as parameter counting, homogeneous models are infinitely improbable, and isotropic models are improbable by a further order of infinity.

The question is, how many of the infinity of infinity of worlds undergo inflation and are isotropized? Only a few have been tested. Roger Penrose of Oxford University, resurrecting an older argument brought against the Mixmaster universe by John Stewart of Cambridge University, has attempted to show that inflation cannot in principle isotropize all possible universes. Penrose imagines a universe that is highly inhomogeneous today. Beginning with these "final" conditions and running the equations of relativity (including inflation) backward to the Big Bang results in a set of initial conditions that are far more inhomogeneous. The inhomogeneities cannot be smoothed out by inflation because they have been chosen specifically to produce an inhomogeneous universe today. (In fact, such a model will usually not inflate at all.) Whether Penrose models outnumber inflationary models remains an open question.

Even before Penrose raised his objections, it was clear that to produce enough inflation in isotropic models there must be "fine tuning" of the initial conditions. In order to get around the fine tuning problem, Andrei D. Linde of the Lebedev Physical Institute in Moscow proposed "chaotic inflation" in 1983. He imagines the universe created with highly inhomogeneous initial conditions. Each region of space is different from every other. In some regions inflation takes place, in others not. In some places inflation is enough to isotropize the universe, in others not. Therefore, on the largest scales, the universe remains highly inhomogeneous even after inflation. But Linde then argues that we happen to be observing a region of the universe that appears flat and isotropic because such regions preferentially permit life to exist.

To many physicists, "anthropic" arguments—arguments based on the existence of life—are anathema. They are nonetheless difficult to avoid when confronting the inflationary scenario. The most famous prediction of inflation is that the universe should be extremely close to the critical density—the density needed to change an ever-expanding FLRW model to a closed FLRW model, which will eventually recollapse. An FLRW model at the critical density is by definition flat. This prediction (as well as claims that inflation smooths out inhomogeneities) clearly depends on the time of observation. Even if inflation succeeds in flattening the universe, the equations of relativity show that it must eventually become curvature dominated again, driving the density far from the critical value. That we observe the density to be near critical is due to the fact that we are observing it at this particular time in the universe's history. Arguing from the anthropic principle, one would maintain that at any time when the density was far from critical, life could not exist and no observations could be made; therefore we must observe the density to be near the critical value.

A perhaps less speculative cure for the observed approximate homogeneity of the universe is to assume the universe is closed. Inflation produced a homogeneous universe by vastly expanding the horizon around any particle; interaction among particles was then postulated to smooth out inhomogeneities. An open universe, however, is infinite in extent, and as an observer's horizon expands it encompasses information from ever larger regions of space. A new region of space may be radically different from the old, and so even if inflation has smoothed out irregularities, eventually an observer may encounter new ones. As the horizon expands in a closed universe, on the other hand, it continually reencompasses the same spatial regions, so there are no surprises, and physical processes can isotropize all parts of the universe.

An extreme version of this viewpoint does not require inflation at all; it merely requires that we live in a closed universe that is actually much smaller than it appears. The "small-universe model," developed by one of us (Ellis), consists of a basic cell in which a few million galaxies are distributed in a highly inhomogeneous fashion—in contradiction to observations. The sides of the basic cell, however, are matched up so that light from each galaxy can travel around the cell many times since the Big Bang. Thus any observer sees the image of a given galaxy many times.

An exactly analogous situation is encountered in a hall of mirrors; one object produces a large number of images. With enough reflections it becomes extremely difficult to identify the multiple images with the original object—in this case, a galaxy. Moreover, the images become distributed more and more homogeneously with each pass, until in the limit of an infinite number of passes the distribution becomes totally random. At some number of passes, an inhomogeneous small universe should be indistinguishable from the real universe.

The small-universe model might be tested by looking for structures, such as filaments and voids, that repeat themselves at regular intervals in the universe. This interval would be the size of the basic cell, which is constrained by current observations to be at least 8 percent of the radius of the observable universe. Put another way, light would have circled the cell about 12 times since the Big Bang. If refined observations push the repetition scale to larger and larger values, the hypothesis becomes less attractive because the small universe becomes large. As it stands, though, the model gives an economical way to create the observed approximate isotropy of the universe and also demonstrates how difficult it is to distinguish one cosmological model from another.

Up to this point it might appear that we have justified inhomogeneous models as much by philosophical arguments as by physical arguments. Recently considerable interest has centered on a model that is naturally, or unavoidably, inhomogeneous. Certain models of particle physics hold that at about 10^{12} degrees Kelvin, neutrons and protons are broken down into their more fundamental constituents, the quarks. Such a temperature would exist at roughly .01 second after the Big Bang. As the universe cooled below this temperature, quarks would combine into neutrons, protons, and mesons, particles collectively known as hadrons. The phase transition would take place unevenly at different points in space, much as the freezing of water begins at one point in a glass and extends to other regions. The result would be an inhomogeneous distribution of hadrons, including baryons (neutrons and protons) throughout the universe; the baryon density would be higher in some regions than in others.

In the standard-model nucleosynthesis calculations, which we described above, the neutron-proton distribution is assumed to be uniform throughout space. One great puzzle in current cosmology is that such nucleosynthesis calculations show that the density of baryons cannot be more than about 10 percent of the "critical density"—the density needed to close the FLRW model, which is predicted by inflation. Higher baryon densities push helium above observational limits. Because inflation is such a popular theory, many attempts have been made to resolve the dilemma.

Several years ago James H. Appelgate, Craig J. Hogan, and Robert J. Scherrer proposed that the inhomogeneities in baryon density produced by the quark-hadron phase transition provide the required resolution. Protons, being positively charged, do not move rapidly, because they are trapped in the surrounding sea of negatively charged electrons. In contrast, neutrons, having no charge, diffuse rapidly from high-baryon-density regions to low-baryon-density regions. The result is that some regions of space contain more neutrons than protons, and vice versa. Since each helium nucleus contains two neutrons and two protons, the amount of helium produced in each region of space depends on the relative numbers of neutrons and protons (Figure 7.3).

Appelgate, Hogan, and Scherrer argued that the redistribution of

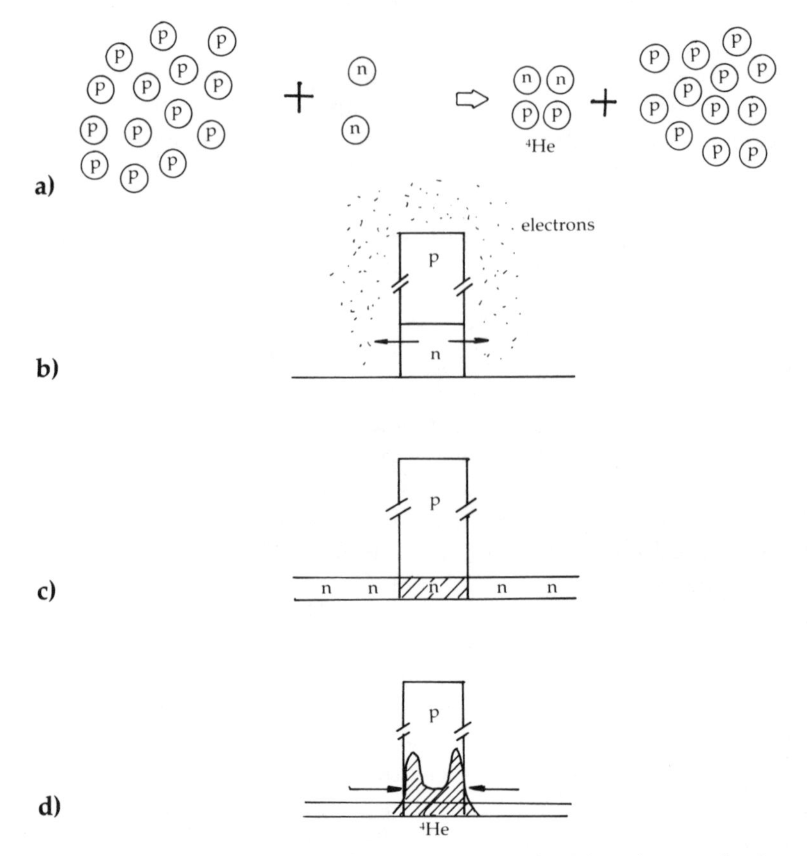

Figure 7.3 (**a**) The amount of helium (^{4}He) produced in the standard model depends on the relative abundances of neutrons and protons at three minutes after the Big Bang. In the standard model protons outnumber neutrons by about seven to one at that time. Since helium requires two neutrons and two protons, this results in about 25% (by mass) of the universe being incorporated into helium and 75% left over in protons.

The helium fraction might be lowered by inhomogeneities in the neutron and proton density. Assume an extreme case in which all the neutrons and protons are concentrated in one region (**b**). Then some of the (electrically uncharged) neutrons will diffuse into the empty regions, while the (charged) protons are held in place by surrounding electrons. The neutrons that diffuse out are lost to the nucleosynthesis process, since there are no protons to combine with. Helium can form only in the shaded region (c), and since this region is neutron-depleted, the final helium abundance is lowered. Unfortunately, in a more realistic model (d), as neutrons are used up in the nucleosynthesis process, the neutrons that diffused out rapidly diffuse back in to equalize their abundance. Helium is formed in the shaded region, with the result that the final helium fraction is not changed significantly from the standard value.

neutrons by diffusion actually lowers the amount of helium produced in nucleosynthesis. Moreover, a simple model developed by them indicated that the average baryon density could be as high as the critical value and helium would still be within observational limits, as would the abundances of deuterium and other trace elements.

More sophisticated numerical computations by Kurki-Suonio, Centrella, Wilson, and two of us (Matzner and Rothman) have shown that for all the parameters investigated, the promising idea does not work. Nucleosynthesis starts first in regions of high density, using up available neutrons. Neutrons that had previously diffused into low-density regions then diffuse back into the high-density regions to equalize the neutron abundance; any irregularities are damped out, and the final helium abundance is not much different from that produced in the standard model. This means that if the baryon density is at the critical value, the final helium abundance is well above observational limits.

The nucleosynthesis results provide a good example of cosmology's persistent problem. For many years cosmologists devoted most of their attention to the exactly isotropic and homogeneous standard model. It has served well in its predictions of the element abundances and the microwave background. Nevertheless, for philosophical and logical consistency, as well as for theoretical and observational reasons, it has become increasingly clear that nonstandard models must be studied.

Unfortunately, it is not clear that the models so far developed explain the observed features of the universe any better than the standard model. There are many nonstandard models that we have not discussed in this article. For instance, quantum gravity, in which quantum mechanics and relativity are wedded, may provide a prediction of initial conditions based on first principles. On the other hand, quantum cosmology is a still-speculative theory necessary only for times not much longer than 10^{-43} second after the Big Bang. Given that inflation is meant to wipe out virtually all initial conditions to give the universe we currently observe, it may be that a theory of quantum cosmology will be forever untestable. Faced with the uniqueness of the universe, such a situation would force the choice between models to be made on a philosophical rather than experimental basis.

CHAPTER 8

Stranger Than Fiction: Cygnus X-3

(WITH DAVID ASCHMAN)

INTRODUCTION. "EVERYTHING WE SAY WILL BE WRONG."

It is not difficult to calculate that if one inflated the world to keep up with the current rate of population growth, then after 2,598 years the earth would be expanding at the speed of light. The growth of science is proceeding even faster. Several years ago, in physics at least, we crossed the point at which the expected lifetime of a theory became less than the lead time for publication in the average scientific journal. Consequently, most theories are born dead on arrival and journals have become useless, except as historical documents.

You may remember, about twenty years ago, the discovery of polywater, which was proven to be erroneous. At about the same time, Joseph Weber announced the detection of gravitational waves, a result that to this day has not been duplicated. The spring of 1980 witnessed the Great Neutrino Scare, a cascade of announcements that the neutrino had, after all, a rest mass. None of these experiments have been successfully duplicated. Since then the flow of irreproducible results has increased dramatically. In the second half of 1985 alone, at least a half dozen newly claimed discoveries in particle physics disappeared and the public never heard of them. As we write this, the Hypercharge Scare—the prediction that objects of different masses do fall at different rates—has apparently evaporated, after a lifetime of only two months.

This state of affairs has made popularization a dangerous activity, because by the time you have researched, written, and published an article, the subject may be ancient history. Nevertheless, occasionally a mystery surfaces that is exciting enough to risk instant obsolescence. One of these mysteries is Cygnus X-3, which has scientists rethinking conventional theories of particle physics. Cygnus X-3, unlike other events of the last year, has managed to survive for at least eight months, which bodes well for the future. We cannot guarantee that by the time you read this Cygnus X-3 will not have been forgotten, but in the meantime it will have generated a lot of thought—and a lot of fun.

CYGNUS X-3. "THIS WILL NOT BE WRONG."

As its name implies, Cygnus X-3 is the third brightest x-ray source in the constellation Cygnus the Swan (Figure 8.1). It happens not to be visible in the optical spectrum, because interstellar gas and dust absorb virtually all the radiation at these wavelengths. But, in addition to the x-ray spectrum, CX3 is observed at infrared, radio, and gamma wavelengths. Now, it is important to understand that x rays are simply photons, like ordinary light rays but of a much higher energy. A typical x-ray photon will have an energy of roughly 10,000 electron volts—10,000 times the energy of an optical photon. The "electron volt" is just a convenient unit of energy for our purposes, like the more familiar "ergs" or "joules." To give you a feel for the numbers involved, the energy inherent in the mass of one neutron (by $E=mc^2$) is about 1 billion eV. This is usually abbreviated 1 GeV, where GeV stands for "giga electron volts," or 10^9 eV. The most powerful accelerators on earth produce particles of 10^{12} eV or 1 TeV (T for "tera"), and the proposed Superconducting Supercollider will generate particles of about 20 TeV.

We said that CX3 is also a gamma-ray emitter. Confusingly, the

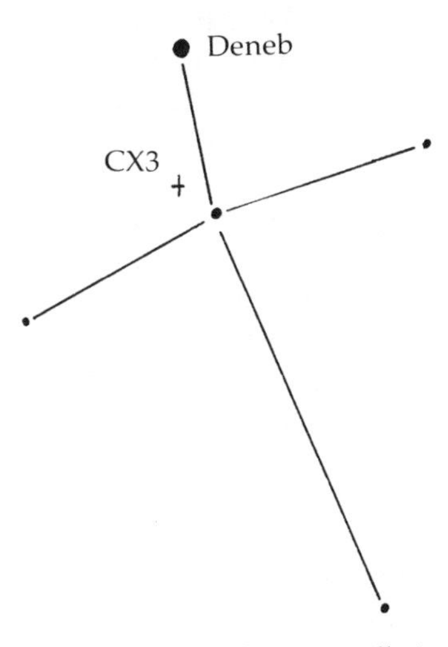

Figure 8.1 X marks the spot. The familiar constellation of Cygnus is shown along with the position of Cygnus X-3.

term "gamma ray" in fact refers to any photon, but astronomers tend to use it to designate the highest-energy photons (those with more energy than x rays). For our purposes we may somewhat arbitrarily call any photon above 10^5 eV a gamma ray. It turns out that not only x rays and gamma rays, but also "cosmic rays" are important for the CX3 story. Cosmic rays are distinguished from gamma rays in that they are not necessarily photons; the term refers to *any* high-energy particles (usually, in fact, protons) that impinge on the earth's atmosphere from outer space. When these primary cosmic rays strike the nuclei in the upper atmosphere, they initiate a complex cascade of reactions known as a cosmic-ray shower. The "secondaries" produced in the shower are what earth-based cosmic-ray detectors actually observe. The energy of cosmic-ray primaries can be as high as 10^6 TeV; this is such a high energy that to explain them has been one of the long-outstanding problems of physics.

As mentioned, CX3 is the third-brightest x-ray source in Cygnus. But this is from the point of view of an observer on earth. CX3 lies at a distance of almost 40,000 light-years from us. Intrinsically, CX3 is brighter than Cygnus X-1, the leading black-hole candidate, whose distance is only about 6,000 light-years. Intrinsically, CX3 is one of the most luminous objects in the galaxy; it is not less than 100,000 times brighter than our own sun. Even more remarkable, much of this luminosity is concentrated in particles with energies of up to 10^6 TeV, one million times more energetic than those produced by the most powerful earthbound accelerator. It is now thought that CX3 and the few other objects like it are responsible for producing the ultra-high-energy cosmic rays just discussed. The mechanism for producing these primaries will be made more clear below.

Until 1983 CX3 was an extremely interesting but well-behaved object. Its most characteristic feature is that the x-ray and gamma flux peaks every 4.8 hours. This "duty cycle" has been documented by numerous observations (including those of the Uhuru x-ray satellite), and there is no doubt that it exists. To explain the periodicity of the emissions is not difficult. CX3 is almost certainly an "eclipsing binary" (Figure 8.2a). A compact object, presumably a neutron star, and an ordinary companion like our sun revolve around each other with a period of 4.8 hours. The neutron star's intense gravitational field siphons gas from the envelope of its companion. As the gas falls toward the neutron star, it is heated by internal friction to temperatures up to 1 TeV. In the process, high-energy x rays and gammas are emitted. When the neutron star passes behind the companion, the companion increasingly absorbs the x-ray and gamma flux until we on the opposite side observe a minimum. When the neutron star passes to our side of the companion, the flux rises to a maximum.

CX3 is not entirely peaceful, however. Every 367 days it flares up, and radio emissions increase a thousandfold. No one knows what causes CX3's outbursts or why they occur periodically. One can imagine a third star in the system, which interacts with the other two in a way that produces a flare every 367 days. In any case, these are very violent events.

The neutron star in CX3 also provides an explanation for the ultra-high-energy cosmic rays, which are too energetic to be produced simply by hot accreting gas. The magnetic field of the neutron star—which is literally trillions of times stronger than the earth's own field—acts as an accelerator and drives electrically charged particles, such as protons, to energies of up to 10^6 TeV. They are then scattered throughout the galaxy, where their trajectories are bent by the galaxy's magnetic field. Eventually, after wandering around in interstellar space, they are observed on earth as cosmic rays coming from all directions. Shortly after leaving the surface of the neutron star, however, some of the accelerated protons strike other protons in the atmosphere of the companion and produce a cascade of debris that contains an elementary particle known as the neutral pion. The pion, in turn, decays into two gammas of about 10^5 TeV, which travel on to earth. Thus the neutron star provides a source for both ordinary cosmic rays and CX3's highest-energy gammas. Note that these gammas (as distinguished from the lower-energy particles emitted by the hot accreting gas) are produced by accelerating a proton through the envelope of the companion star. The two-stage process is necessary because gammas are electrically neutral, and so cannot be accelerated by either an electric or magnetic field. It is fairly clear (Figure 8.2b) that you would expect to see the ultra-high-energy particles only when the neutron star is just grazing the companion. When the neutron star is between us and the companion, the protons cannot interact with the companion's atmosphere and hence will not produce the pions that subsequently decay into gammas. When the neutron star is directly behind the companion, the high-energy gammas are eclipsed, as is radiation at other wavelengths. Consequently, you predict that the ultra-high-energy photons from CX3 should appear twice every 4.8 hours. Such behavior is indeed observed.

If this were the entire CX3 story, it would be fascinating enough, but the excitement would be largely limited to the astronomical community. The fever spilled over in mid-1985 with the announcement of some recent results. Specifically, in 1983, deep underground at the

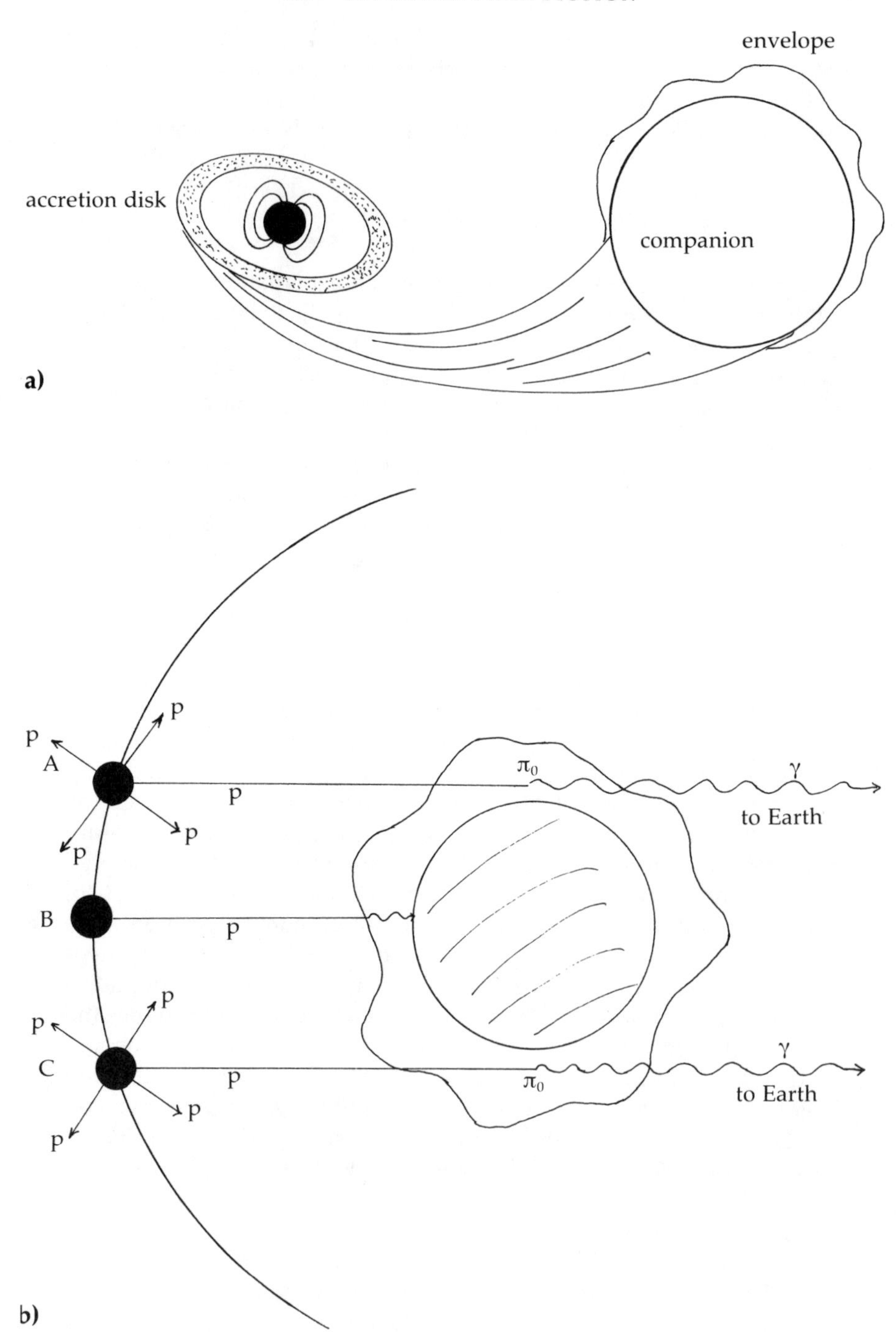
envelope
accretion disk
companion
a)
p
p
A
p
p
p
B
p
p
p
C
p
p
p
π_0
γ
to Earth
π_0
γ
to Earth
b)

Figure 8.2 (**a**) The generally accepted picture of Cygnus X-3. The intense gravitational field of a compact object (black) siphons off gas from the envelope of its ordinary companion. As it falls toward the compact object, usually assumed to be a neutron star, internal friction heats the gas to millions of degrees, until it emits x rays and gamma radiation. This radiation is eclipsed every 4.8 hours when the neutron star passes behind the companion from our point of view. Close to the neutron star, the gas forms a rotating accretion disk. Eventually, the same collisions that heated the gas particles slow them down so much that they fall out of the disk and onto the neutron star.
This mechanism cannot explain the highest-energy gammas from CX3. The rotation of the disk causes particles (say, protons) to move at right angles to the magnetic field. The protons therefore experience an acceleration directed radially outward from the neutron star. The protons (p) will fly off in all directions (**b**). Some will pass throught the atmosphere of the companion star where they will produce a pion, which in turn decays into the gammas (wavy lines) that are eventually observed on earth. The two-step process is necessary because gammas are neutral and cannot by themselves be accelerated at all, let alone to 10^5 TeV. Note also that we will observe these gammas only when the neutron star is grazing the companion (positions A and C). At position B, the companion eclipses any radiation given off by the neutron star. When the neutron star is on our side of the companion (not shown), any proton directed toward us will not interact with the companion's atmosphere and hence will not produce a gamma.

Soudan iron mine in Minnesota, teams from the University of Minnesota and Argonne National Laboratory failed to observe the decay of the proton.

As you may know, the new Grand Unified Theories predict that the proton should decay with a lifetime of approximately 10^{32} years. This means that if you collected 10^{32} protons, you would be able to observe, on average, one decay per year. A thousand metric tons of water—roughly a large swimming pool's worth—contains about this number of protons. The Soudan detector is just such a swimming pool, surrounded on all sides by photomultiplier tubes to catch any light given off by decaying protons. To the great dismay of physicists, however, neither the Soudan detector nor others like it, despite several years of operation, have observed the expected events.

What has happened is something very unexpected. In 1983 the Minnesota/Argonne research group began to observe *muons* in the Soudan device. Muons are common elementary particles, about 200 times more massive than electrons, but otherwise with all the same properties. ("The muon? Who ordered that?"—I. I. Rabi, Nobel laureate.) When a muon enters the swimming pool, it is traveling faster than the speed of light in water (which is only two-thirds the speed of light in a vacuum) and therefore emits "Cherenkov radiation" in a narrow cone along its trajectory (Figure 8.3). This cone of blue light will trigger a ring of photomultipliers on the swimming pool wall. If the muon comes from a different direction, a different set of photomultipliers will be set off. Thus by examining which phototubes are triggered, the investigator can reconstruct the path of the muon. When this is done it is found, not surprisingly, that muons enter the tank from many directions—more coming down than going up, sim-

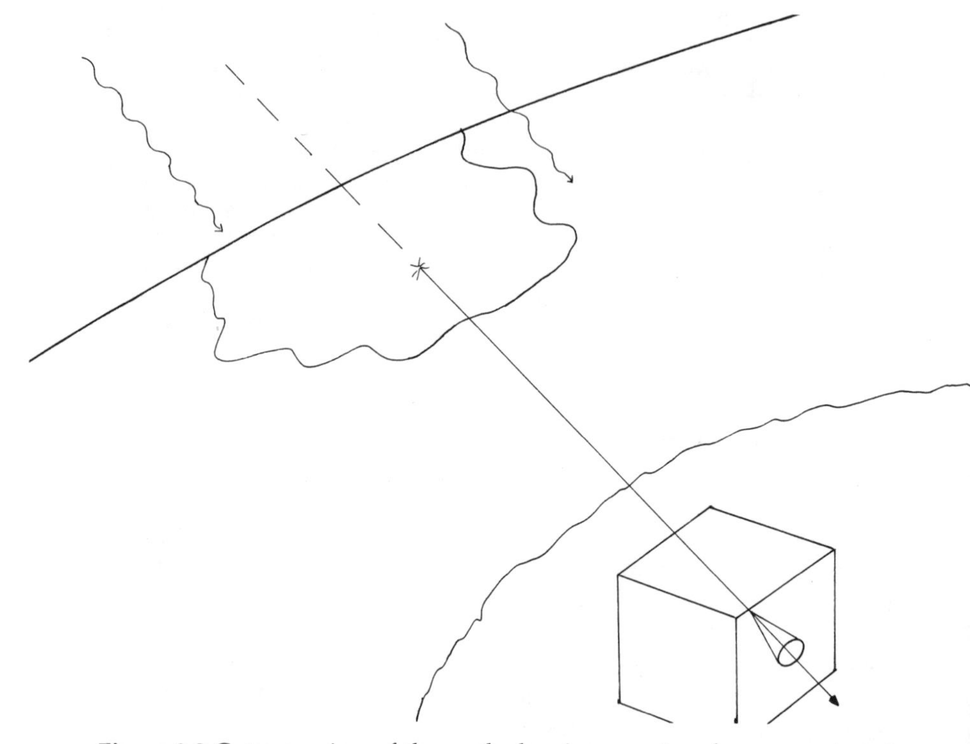

Figure 8.3 Cutaway view of the earth showing a proton decay experiment, which is typically located half a kilometer or so underground. Many particles streak toward the earth, including high-gammas (wavy lines) and the unidentified cygnet (dashed line). The cygnet collides with nuclei in the earth's crust and produces muons (solid line), which continue toward the detector—a large tank of water. Because the muon's velocity is greater than the speed of light in water, it emits "Cherenkov radiation," which is the familiar blue light seen in photographs of nuclear reactors. This cone of Cherenkov light strikes the photomultiplier tubes that line the walls of the tank. By tracing back along the cone, one can approximately reconstruct the path of the muon.

ply because the bulk of the earth lies beneath the detector and acts as an effective shield against muons attempting to pass through. This nearly uniform muon flux is known as the background. But in the Soudan detector, scientists also observed a significant excess of muons coming from one particular direction in the sky. This direction was a box of about 3° × 3° centered on Cygnus X-3. What is more, the flux peaked every 4.8 hours. And finally, the arrival of these muons did not lead or lag the periodic x-ray and gamma signal, but remained exactly in phase.

For reasons that will become clear, scientists were—and many remain—properly skeptical of the claims. But they cannot be dismissed entirely out of hand. The European group NUSEX (Nucleon Stability Experiment), which operates a similar detector in the Mont Blanc tunnel, has apparently discovered the same effect. Furthermore, an earlier experiment by the University of Kiel group showed that in cosmic-ray showers from the general direction of CX3, the muon content was roughly ten times the muon content of cosmic-ray showers originating in other parts of the sky.

We are not yet done. Between October 3 and October 13, 1985, the number of excess muons above the background entering the Soudan detector jumped by a factor of about ten. October 12 coincided with CX3's 367-day outburst, which was the largest yet recorded. Finally, on the same night of October 12, a cosmic-ray detector located atop the Haleakala volcano in Hawaii (and run jointly by Purdue University, the University of Wisconsin, and the University of Hawaii) made a spectacular observation: cosmic-ray showers from the direction of CX3 inundated the device for 60 seconds and then suddenly switched off. The usual cosmic-ray shower lasts about one-billionth of a second. This one lasted for a full minute and had an intensity significantly larger than background—by four standard deviations. The showers also came at nearly the same point within the 4.8-hour period as the muons observed at Soudan.

Before explaining why the CX3 observations are so strange, and exciting, we should perhaps inject a word of caution. We have described the muon signals as large and significant, but the actual number of muons involved is in fact rather small. The Soudan group's published results contain only 64 (84 ± 20) muons from the direction of CX3 above the background to substantiate the 4.8-hour period and only 20 events to coincide with the October outburst. The NUSEX data sample is even smaller, with 13 (19 ± 6) events to substantiate their claims. Anyone who has worked with the statistics of small numbers knows that the removal of one or two data points can cause a result to vanish. Indeed, another collaborative effort, the Irvine-Michigan-Brookhaven (IMB) group, has not reported the 4.8-hour signal in their detector, though they are now looking hard through the data. And Francis Halzen, a physicist from the University of Wisconsin who works on the CX3 problem, has said, "My gut feeling is that the signals are spurious in some way we haven't understood." But others have estimated that the odds that the Soudan and NUSEX observations are due to chance are less than one in a thousand.

The most peculiar aspect of CX3 does not lie in the astrophysics, as peculiar as the astrophysics may seem. If the Soudan and NUSEX results are correct, our present concept of particle physics may have to change.

The origin of the problem lies in the fact that both the Soudan and NUSEX devices are detecting muons. Muons are unstable particles and very short-lived. At rest they decay with a lifetime of 2.2 microseconds. This lifetime can be extended by boosting the muons to very high energies, at which relativistic time-dilation effects become important. Now, the muons observed at Soudan have an energy of about 1 TeV and are thus traveling at 0.99999995 times the speed of light. Time dilation increases their life by a factor of about 3,000, to 6 milliseconds. But this is still a (very) far cry from the 40,000 years needed to travel to earth from CX3. All the muons would have decayed before they left the CX3 system. Thus the muons must be secondary particles originating in the earth's atmosphere or crust. What are the long-lived primary particles? This is the question that has scientists all over the world scratching their heads in puzzlement. No known particle seems to fit the bill. The mysterious primary has appropriately been dubbed the "cygnet," and speculation regarding its identity has been, if not immediately believable, at least highly imaginative.

CONVENTIONAL CYGNETS. "NO GO."

The cygnet cannot be electrically charged. This is so because the galaxy's magnetic field deflects any charged particle from its original flight path. For instance, a 10^6 TeV proton emitted by CX3 would have its trajectory bent into a complete circle by the time it had traveled only one-thirtieth the distance to earth. Consequently, we would never observe the secondary muons coming from the direction of CX3. So the cygnet must be neutral.

The neutron is the first obvious candidate. However, like the muon, the free neutron is unstable and decays with a lifetime of about fifteen minutes. Just as for muons, boosting the neutron to ultra-high energies will extend its lifetime. To enable a neutron to travel 40,000 light years before decaying, you must give it 10^6 TeV of energy, 10^9 times its rest energy of 1 GeV. Then the neutron might reach earth, collide with molecules in the air or rock, and, through known reactions, produce the muons that are observed underground.

The proposed solution does not work. First of all, 10^6 TeV is at the upper limit of energies produced by CX3 and other cosmic-ray sources. If one considers *all* the known cosmic rays of this energy, the total flux would not be enough to produce more than one muon per year in the Soudan detector. There is a more dramatic way of saying much the same thing. If you assume that each neutron is responsible for the subsequent production of one muon, you quickly calculate that CX3 must emit 10^{42} ergs per second in neutrons. This is the luminosity of a small galaxy. Anything emitting 10^{42} ergs per second in neutrons would have to emit comparative amounts of radiation in other forms—like gammas—so we would observe a small galaxy sitting in the Milky Way. Presumably this can be ruled out. On the other hand, recall that the observed muons have an energy of roughly 1 TeV—one-millionth the energy of the hypothetical neutrons. You might assume that each neutron could give rise to one million muons. In this case, CX3 would need to emit only 10^{36} ergs per second in neutrons—1,000 times the luminosity of the sun. Such an alternative is ruled out simply because neutrons do not produce one million muons. A neutron at 10^6 TeV initiating a shower in the earth's atmosphere or crust will produce only several hundred muons, maximum.

The second candidate for the cygnet is the neutrino. Now, the neutrino has a very small interaction cross section, which is the technical way of saying that it collides with other particles only very rarely. The usual illustration is that the average neutrino will travel through four light-years of lead before suffering a collision. In the Soudan detector, on the other hand, the muon signal is observed to fade out as CX3 approaches the horizon. This behavior is easily interpreted: Muons produced in showers by heavy particles such as protons are created near the surface of the earth, since the proton has a large cross section and typically interacts in the first few hundred meters of rock or earlier. As CX3 sets each night, muons created in such showers must travel through ever greater amounts of rock in order to reach the detector. The longer the path length, the smaller the number of muons that reach the detector and are registered—exactly as observed. But neutrinos do not particularly care whether they travel through one kilometer of rock or one thousand. If neutrinos were the primaries they would be just as likely to create muons near the detector as far from it. Thus the signals would not fade out as CX3 approached the horizon.

The last obvious candidate for the cygnet is a high-energy photon or gamma ray. The argument for ruling out the photon assumes that we know all the possible reactions by which a photon produces a muon. Then calculations show that a photon-initiated shower generates too few muons by a factor of 300–1000. Unless there is something quite new happening—a dramatic increase in the probability of photons interacting with matter at very high energies—gamma rays are excluded.

That ends the list of obvious candidates from among the elementary particles. Thus, barring mistakes, the simplest explanations all fail. At the next level of complexity you might wonder whether the role of cygnet could be taken by a composite neutral particle, such as a carbon atom. Unfortunately, this suggestion is also untenable. It is, first, not very likely that CX3 emits carbon or other elements. At any rate, recall that the muon signals have a 4.8-hour period. Furthermore, all the muons arrive within about six minutes of each other. Now if the cygnet is massive, the velocity depends on the energy of the particle and on the mass. If the cygnets were emitted with even slightly varying amounts of energy, they would travel to earth at different velocities and arrive at different times. The six-minute bunching of arrival times would thus be washed out. If we assume a spread in energy of ten between the lowest- and highest-energy cygnets, we can easily calculate that the upper limit on the cygnet mass is roughly 2 GeV, or about twice the proton mass. Hence more massive particles, such as atoms, are excluded.

EXOTIC CYGNETS. "MAYBE."

Physicists have now exhausted conventional explanations for Cygnus X-3. We reiterate that mistakes often occur. Some years ago one of us (T. R.) was observing the radio source Cassiopeia-A at the National Radio Astronomy Observatory and detected an exceedingly curious signal. This momentous discovery later turned out to be the sun. Other radio astronomers have been deceived by rush-hour traffic and farm-tractor spark plugs. It is conceivable, though unlikely, that the Soudan and NUSEX signals are similarly mundane. Mundane explanations are often as elusive as exotic ones, and not as exciting. So scientists have put the mundane behind them and untethered their imaginations. We now enter the realm of speculation.

The most spectacular suggestion, and one that has already been advanced by a number of groups, is that the neutron star in CX3 is not a neutron star at all—it is a quark star. The cygnets themselves are quark matter that has been detached from the surface of the star and somehow accelerated toward earth. (We return to possible mechanisms below.) As you probably know, current theories suggest that all heavy particles such as neutrons and protons are composed of the more fundamental quarks. (We should emphasize that this theory is more than speculation; to the contrary, it is strongly supported by all experiments to date.) The standard version of this theory calls for six quarks, somewhat whimsically termed "up," "down," "strange," "charmed," "bottom," and "top" (udscbt). The proton is composed of two u's and a d, while the neutron contains two d's and a u.

The fact that quarks combine to form neutrons and protons, but neutrons and protons do not disintegrate into free quarks, indicates that ordinary matter (neutrons and protons) is more stable than quark matter. Otherwise we would all long ago have disintegrated into quark soup. However, this statement is true only when we speak of the up and down quarks that constitute ordinary matter. The situation may change when strange quarks are added. In this case it turns out that at high enough densities—like the densities found in neutron stars—"strange quark matter" may be more stable than ordinary matter. (The reason for this has to do with the fact that quarks are so-called "fermions." No two of them can occupy the same quantum state.) This interesting possibility allows for the existence of quark stars.

Quark stars were postulated ten years ago by a number of researchers. At that time, however, there was no experimental motivation for the suggestion. The CX3 muons have provided just such a motivation. If you believe CX3 harbors a quark star, several candidates for the cygnet immediately suggest themselves.

In a recent *Physics Letters* paper, Gordon Baym, Edward Kolb, Larry McLerran, T. P. Walker, and R. L. Jaffe propose that the cygnet is the "doubly strange dihyperon" or "dilambda," a particle first dreamed up by Jaffe in 1977 but so far not observed. The dilambda consists of two up quarks, two down quarks, and two strange quarks. It is neutral, which is required of the cygnet, and fairly light, with a predicted rest mass of about 2 GeV. Furthermore, the dilambda may be stable enough to last the 40,000-year flight. For the moment, then, the dilambda appears to be a good candidate for cygnet.

It is not the only candidate; other suggestions have been made, for instance, stable "bags" containing nearly equal amounts of u, d, and s quarks. But such suggestions are variations on the dilambda theme. The basic idea is that the source of the cygnet must be a quark star.

But the supposition that CX3 contains such an object brings with it a number of difficulties. Originally, quark stars were believed to be essentially neutron stars with a quark core. If this is the case, it is very difficult to imagine how to get the strange matter out. First, if quark matter exists only in the deep interior of a neutron star, then you must find a mechanism—such as diffusion—to bring it near the surface. The surface of a neutron star is itself believed to be an ultrahard crust of ordinary nuclear matter. So, to excavate the strange matter, you must postulate something like a dilambda geyser (Figure 8.4) that erupts continuously, so that every 4.8 hours, when the quark star passes from behind its companion, we get a burst of cygnets. Or perhaps instead of a geyser there exists a "strange quark matter volcano" (it is difficult to resist: "Quarkatoa"). Each of these scenarios is very cumbersome, and they do not enjoy much favor among scientists.

Baym *et al.*, in their paper in *Physics Letters*, propose a more credible model. Imagine the detonation of a supernova. The envelope of the original star is blown away, and the core collapses into a neutron star. If the density of the neutron star is high enough, at least in certain regions, then in these pockets ordinary matter will be transformed into more stable quark matter. Until this point the scenario is identical to that just discussed. But because the strange matter pockets are more stable than their surroundings, Baym *et al.* suggest that they will begin to "gobble up" the regions of ordinary matter until the entire neutron star is converted into a quark star. In this picture, quark matter exists all the way to the surface of the compact object.

Perhaps you find such a scenario hard to swallow, in which case you are not alone. But if you do buy it, it becomes less difficult to get the quark matter off the surface of the star. Because the accreting gas from the ordinary companion is highly energetic by the time it hits the quark star, it could simply eject blobs of strange matter as it strikes the surface. The ultra-high magnetic field of the quark star accelerates the electrically charged blob of strange matter from the surface—exactly as it did for cosmic-ray protons. If this blob passes through the atmosphere of the companion, it will disintegrate into smaller fragments. One of these could be the dilambda, which travels on to earth. Again, you need a two-stage process, because dilambdas are neutral and cannot be accelerated.

Admittedly, this is all rather vague, since no one has yet worked

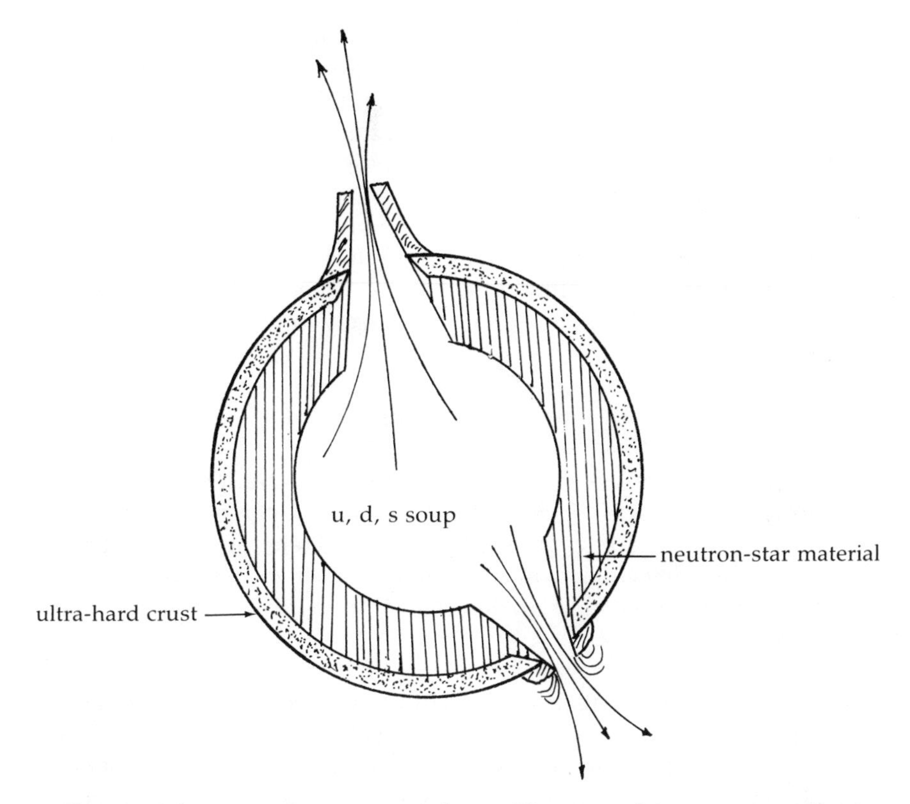

Figure 8.4 A proposed neutron-quark star. The crust of a neutron star (dots) is thought to be an ultra-hard lattice of nuclear matter. Beneath that is ordinary neutron-star material, mostly neutrons (shade), with the center of the star, in this model, composed of a free-quark broth. Unfortunately, it is difficult to imagine how to get quark matter to the surface and out. One must postulate a "quark geyser" or "strange matter volcano." Such suggestions are not taken too seriously. More likely, the quark matter exists all the way to the surface of the star and is ejected as the infalling hot gas strikes the surface at high velocity.

out the details of blob emission and conversion into dilambdas. More detailed calculations are needed to firm up these conjectures. But already it is clear that there are some problems with these models. For instance, we mentioned that the Soudan muons appear to come from a region 3° × 3° centered on CX3. If the cygnets are neutral and travel in exactly straight-line paths to earth, why do they not appear to come *exactly* from CX3? The resolving power of the detectors is better than 3° × 3°, so it does not seem to be an instrumental effect. The models we have been discussing do not explain this observation.

This, to date, is the story of Cygnus X-3. We cannot predict with certainty the outcome of the current investigations. Regardless of the eventual result, Cygnus X-3 will remain a tantalizing object. Scientists are attracted to problems of subtlety, more so if they contain a trace of the bizarre. Francis Bacon put it this way: "There is no excellent beauty that hath not some strangeness in the proportion." And so, if the Soudan and NUSEX signals turn out to be spurious, we merely learn once again that there are many more ways to get something wrong than to get something right. If the signals turn out to be genuine, we learn again that the universe is queerer than we can suppose—a conclusion that is, in any case, likely to remain.

POSTSCRIPT

The tale of Cygnus X-3 has an O'Henryesque ending. "Stranger Than Fiction" was written in 1985. Shortly thereafter Cygnus X-3 vanished from the scientific scene: groups running the Frejus and Kamiokande proton-decay detectors, which are more sensitive than the Soudan and NUSEX devices, failed to detect any muon excess in the direction of CX3.

For two years there was only silence. But then in 1988 a large research group consisting of workers from the University of Maryland, George Mason University, Los Alamos National Laboratory, and the University of California, Irvine, reported the results of observations taken with cosmic-ray detectors: there appeared to be an anomalously large muon flux coming from the direction of Hercules X-1.

Like Cygnus X-3, Hercules X-1 is a bright x-ray source. Like Cygnus X-3, Hercules X-1 is a compact binary system, probably consisting of a neutron star with a normal companion. The research group also reported a possible muon excess from Cygnus X-3 itself, but the evidence was not so compelling.

The new results are independent of the old, but one can imagine that the researchers involved in the NUSEX and Soudan experiments may be feeling somewhat vindicated. In any case, the question remains the same: what are the cygnets—or perhaps now (we would hesitate to say it), the Sons of Hercules?

CHAPTER 9

The Ultimate Collider

ANTONI AKAHITO
(TONY ROTHMAN)

SINCE the cave dwellers first collided flint against metal to produce fire, natural philosophers have had to resort to ever higher energies in their quest to unlock nature's tiniest secrets. The Superconducting Supercollider is the latest instrument to be brought to bear in the physicist's eternal search for truth, but with the site allocation already taken care of, it is long past time to look toward the next step in mankind's greatest journey. Since one can expect Congressional fluctuations to obstruct the progress of science for some years to come, this month I wish to encourage proponents of small-scale science—amateurs in particular—to grasp the torch and construct a Planck-mass accelerator.

The Planck mass, or Planck energy (the equivalence of mass and energy by $E = mc^2$ makes the terms interchangeable), is the largest energy that physics as now constituted can deal with in any sensible fashion. It is the energy an average particle had 10^{-43} second after the Big Bang when the forces of quantum mechanics and gravity are thought to have been unified. A Theory of Everything, of which current superstring theories may be dim precursors, would explain the unification and in fact could be tested by a Planck-mass accelerator. In principle, there is little difference between such a machine and its ancestors: protons or electrons are accelerated up to Planck energies and collided head on. During the collisions, the projectiles convert their energy into Planck-mass particles, particles that existed at the earliest instants of creation. Boiled down to its essentials, a Plank-mass accelerator simulates the Big Bang.

Beside such a machine, existing accelerators pale into insignificance. Consider the proton. Its mass is about 10^{-24} gram, orders of magnitude below the sensitivity of the best laboratory balances. Through $E = mc^2$ it harbors an equivalent energy of about one billion electron volts (or one giga electron volt, abbreviated GeV). The world's largest accelerator, Fermilab's Tevatron, can accelerate protons to energies of 2,000 GeV, now usually abbreviated as 2 TeV for

2 tera electron volts. The Tevatron, then, can impart to protons about 2,000 times their rest mass in energy. If two such protons are collided together in the Tevatron, this energy can be used to create new particles with masses of approximately 10^{-21} gram, still far below the sensitivity of any laboratory balance. The Superconducting Supercollider (SSC) is designed for 20-TeV operation, only 10 times higher than Tevatron energies.

Cosmic rays do somewhat better: the highest-energy cosmic rays, thought to be produced by astronomical objects such as Cygnus X-3, are measured at roughly 10^6 TeV. Particles produced in cosmic-ray collisions would weigh in at about 10^{-15} gram.

However, grand unified theories (GUTs), which claim to combine the strong, weak, and electromagnetic forces into one strong-electroweak force, are thought to begin operating at about 10^{12} TeV. This is a million times higher than the most energetic cosmic rays, and 100 billion times higher than the expected attainments of the SSC.

Still, we are going for the Max. The Planck energy, where the Theory of Everything is presumed to come into play, corresponds to approximately 10^{16} TeV. This is 10 billion times more energetic than the most energetic cosmic ray. It is 1,000 trillion times more energetic than the particles that will be produced by the SSC. It corresponds to a mass of about 10^{-5} gram, which can be measured on today's laboratory balances.

The first and most difficult step in building the Planck-mass accelerator is finding a name for it. The Superconducting Supercollider has already overburdened the growing list of endeavors anointed with the adjective "super" (now elevated to the rank of nonhyphenated prefix): superconductors, supersymmetry, superparticles, superstrings, supercomputing centers, and Supertuesdays. The Superconducting Supercollider has even managed to usurp two supers in as many words and has acquired a three-initial abbreviation in the bargain.

It is certainly easy to be sympathetic: the SSC will be 87 kilometers in circumference and will cost $8 billion (barring further overruns). Still, all this is vaguely unsatisfactory. "Super" in the context of accelerators has very much the same ring as the term "postmodern" in literature. What does "postmodern" become after ten years? If an accelerator designed for 20 TeV is to be anointed with the adjective "super," then in the next generation we shall be forced to go to "Hypercollider," and then no doubt to "Superhypercollider" and "Hypercolossalcollider," at which point accelerator naming begins to sound like *Sneak Previews*. On the other hand, without superlatives, where would America be? Still, what is super today is superfluous tomorrow. For the Planck-mass accelerator I therefore suggest "Ultimate Collider," or UC. Modest though two initials may be to describe a machine of 10^{16} TeV, it will have to do; as I have said, according to present conceptions of space and time, it does not make any sense to talk about anything larger.

The second step in designing the UC is to consider what sort of power source will be needed to accelerate protons or electrons up to the Planck energy and create Planck-mass particles. A simple arithmetic calculation reveals the first obstacle: the entire energy of a one-megaton atomic bomb converted to planckons (as I shall call them) will produce about three million. Three million planckons may seem like a lot, but it is negligible compared to the beam intensities achieved by today's accelerators. Machines like Fermilab's Tevatron are typically capable of delivering 10^{12} to 10^{13} particles per second to the target. Consequently, to achieve today's beam intensities, the amateur will need the energy equivalent of roughly one million megaton bombs exploding per second.

This computation assumes, of course, that 100 percent of the energy of the atomic bomb goes into making planckons, which is overly optimistic. The actual efficiency of present-day accelerators is difficult to judge. A beam intensity of 10^{13} particles per second at an energy of 20 TeV represents a power of about 30 megawatts. If, as planned, a 300-megawatt power plant will be built for the SSC, a 10^{13}-particle-per-second beam intensity implies an efficiency of 10 percent; the rest is lost to refrigeration of the magnets, transmission lines, and so on. If the beam intensity is only 10^{12} particles per second, the SSC will be about 1 percent efficient. Of course, with room-temperature superconductors (which the do-it-yourselfer can fashion empirically in the kitchen), refrigeration losses can be reduced considerably.

Nevertheless, I want to be on the safe side, and so I shall assume that the prototype UC will have an efficiency of only 1 percent. With a 1 percent efficiency, the power source for the UC will have to provide the equivalent of 100 million one-megaton bombs per second during operations. This is far above the megatonnage available in today's arsenals.

It does correspond, however, to approximately 4×10^{30} ergs per second, or only about a thousandth the luminosity of the sun, which is well within the range of a science-oriented society. The amateur, then, should begin by placing a system of solar collectors in orbit around the sun. If they are placed at the radius of Mercury's orbit, the combined collection area should be at least 4×10^{13} square kilometers, about 660 times the surface area of Jupiter. The solar energy should then be transformed into microwaves, for example, and beamed to the accelerator proper. A large capacitor bank is recommended, for it will significantly reduce the required collection area. (Canal Street in Manhattan has traditionally been a good place to shop for junk parts.)

Having solved the problem of energy supply, the next task is to look into the design of the accelerator itself. Today's machines are predominantly of two types: linear accelerators, or linacs, and synchrotrons. As its name implies, a linear accelerator accelerates particles along a straight line. The world's largest linac is the Stanford Linear Accelerator—universally known as SLAC—with a length of three kilometers. The way a linac accelerates electrons, say, is fairly straightforward. A high-frequency alternating electric field, usually at about 1,000 megahertz, is passed down a microwave guide. The phase of the field is arranged so that it will push the electrons down the cavity. In other words, the electron is accelerated by getting it to ride the crest of a wave. A linac has the disadvantage that it can only accelerate a particle once—from beginning to end. The final energy of the particle is limited by the amount of energy the accelerator can impart to it on one pass.

By contrast, a synchrotron accelerates particles repeatedly around a single circular track. Synchrotrons are thus capable of much higher energies for a given length than the linear accelerator, and largely for this reason the SSC has been designed as a synchrotron. It will also utilize an increasingly popular technique known as "colliding beams," which is why "collider" follows the second "super" in SSC. According to relativity, the energy available to create new particles is much greater when two protons or two electrons collide head-on than it is when they hit a stationary target in the laboratory. A proton collider therefore circulates two beams of protons in opposite directions until they attain the required energy, then forces them into a head-on collision. In the SSC the full 40 TeV of the two protons is then available to create new particles, each of 20 TeV. For a non-colliding-beam synchrotron to produce a pair of 20-TeV particles from a single proton smashing into a laboratory target, it would have to accelerate the proton to an energy of 800,000 TeV.

For this reason, colliding-beam synchrotrons are now considered the wave of the future. Unfortunately, simple considerations show that synchrotrons, whether stationary-target or colliding-beam, cannot be the basis of the Ultimate Collider (without significant difficulties); the amateur is urged to avoid them.

According to a century-old result of Maxwell's theory of electromagnetism, any accelerating charged particle radiates energy. One of the basic problems any accelerator designer faces is knowing how much energy the electrons will lose as they hurtle down SLAC's vacuum chamber, or how much energy protons will give off as they circulate in the SSC's storage rings. Left to themselves, these circulating protons would eventually radiate away all their energy and stop. So some fraction of the energy input of an accelerator simply goes into replacing the energy the particles lose as they are accelerated.

The amount of energy lost in an accelerator depends very crucially on the design. Synchrotrons are prey to an illness appropriately termed synchrotron radiation, the radiation emitted by any charged particle in a circular orbit. In Cornell's 10-GeV synchrotron, a 10.5-MeV boost is given to an electron on each turn, but the losses from synchrotron radiation on each turn are about 8.85 MeV. And so you see that at high energies most of the energy goes not into accelerating particles but into making up radiation losses. Unhappily, synchrotron radiation goes up as $(E/m)^4$, where E is the particle energy and m is its rest mass—in other words, very rapidly. By the time you reached only 10^4 TeV—5,000 SSC energies—an electron circulating in a synchrotron of radius 100 kilometers would be radiating away a Planck mass of energy on every turn.

Radiation losses are, however, inversely proportional to the radius of the accelerator; an obvious strategy, then, is to make the radius of the accelerator larger. This is not very feasible. The radius necessary to keep a Planck-energy electron radiating at less than a Planck energy per turn is roughly 10^{27} times the size of the observable universe.

Because synchrotron radiation losses go as $(E/m)^4$, such losses are less severe for protons, which are much heavier than electrons. Specifically, the proton is almost 2,000 times more massive than the electron, and so at a fixed energy synchrotron radiation losses are about 10^{13} times less. But the factor of E^4 means that once a proton is accelerated to an energy 2,000 times higher than that of an electron, radiation losses will be the same: in an accelerator with a radius of 100 kilometers, at about 10^7 TeV radiation losses exceed one Planck mass per turn. To keep the radiation losses from Planck-energy protons within acceptable bounds, one would need to construct a synchrotron with a radius 10^{14} times the size of the observable universe.

Luckily, there is a solution to this problem. Radiation losses in a linear accelerator turn out to be vastly less severe than those associated with synchrotrons. In a linac the power lost to radiation can always be kept below the input power simply by keeping the energy given to the electrons below the order of 10^6 TeV per centimeter. SLAC provides an "energy gradient" of roughly 10^{-7} TeV per centimeter, which is 13 orders of magnitude below the upper bound. Protons, because they are heavier, are again subjected to a less stringent limit, in this case about 10^{13} TeV per centimeter.

So, there we are. To be conservative, the Ultimate Collider prototype should be constructed as a linac. As long as we keep the energy gradient below the limit of 10^{6} TeV per centimeter, we can attain arbitrarily high energies. Furthermore, we want to make it a colliding linac—two linear accelerators run in opposite directions—to capitalize on the full energy available in head-on collisions.

The first obvious feasibility test is to scale up SLAC to Planck energies. At 10^{-7} TeV per centimeter, however, this calls for an accelerator 100,000 light-years long, somewhat above the size of the galaxy. A collider would be twice as long, with the laboratory area presumably located at the center. Such unwieldy proportions make data collection inconvenient; an experimenter, after throwing the "on" switch, would have to wait 200,000 years for the results.

Again we are saved by the fact that radiation losses in linacs are so small. If we choose to construct a machine with an energy gradient of 100 TeV per centimeter—still far below the limit of 10^{6} TeV per centimeter—the length of the UC is reduced enough so that it would fit within the orbit of Pluto. At first glance an energy gradient of 100 TeV per centimeter, which is one billion times as large as the gradient at SLAC, strikes one as large, if not impossible. SLAC's accelerating field is produced by a bank of over 200 high-frequency oscillators known as klystrons, and they produce about the maximum gradient attainable by conventional methods. The thought of increasing 200 klystrons to 200 billion does not seem very fruitful.

But the klystron is not the only device capable of generating large amounts of power. The highest power available today actually comes from lasers. Even a commercially available CO_2 laser can result in gradients 10 times as high as those at SLAC, and the HELIOS laser at Los Alamos is already up by a factor of 1,000. Some investigators are now talking about future laser-driven machines with gradients of 10^{-4} TeV per centimeter, only a million times less than our goal.

But at this stage two serious technical obstacles must be overcome. First, laser beams are "transverse": their electric field oscillates in a plane perpendicular to the direction of the beam. One needs to invent a scheme that will somehow transform the high electric fields available in lasers into longitudinal motion, which can accelerate particles.

A free-electron laser is an example of a device that does exactly the opposite: instead of converting electromagnetic energy to the motion of electrons, it converts the motion of electrons into light waves (see "The Free Electron Laser," by Henry P. Freund and R. K. Parker,

Scientific American, April 1989). The trick in the free-electron laser is to change the electron trajectories to the desired direction by applying a magnetic field. An accelerator operating by this principle would essentially be a free-electron laser run in reverse. Unfortunately, as soon as you begin to bend electron trajectories with magnets, the problem of synchrotron radiation eventually defeats you. An inverse-free-electron laser will not provide the basis for the UC.

The second obstacle encountered in going to high fields arises from the fact that electrons are bound to atoms with energies of about 10 electron volts. Atomic dimensions are of order 10^{-8} centimeters. One therefore expects that field gradients larger than, say, 10 eV per 10^{-8} centimeters will tear electrons from their nuclei. In terms of our units this upper limit is about 10^{-3} TeV per centimeter, about 100 times larger than the SLAC gradient. Even smaller values—values too small for the UC by a factor of 100,000 or one million—would undoubtedly cause serious damage to the accelerator support structure.

The amateur must therefore look to new media to construct the Ultimate Collider. One promising device is a laser-plasma accelerator (see "Plasma Particle Accelerators," by John Dawson, *Scientific American,* March 1989). In plasmas, or highly ionized gases, electrons are already detached from their nuclei, and so they cannot be further disrupted. Accelerating fields in experimental "beat-wave" accelerators have already reached 10^{-5} TeV per centimeter; 10^{-2} TeV per centimeter is theoretically possible with currently attainable plasma densities, which are obtained from hydrogen gas. This is only 10,000 times below what is required for the prototype UC. Going to denser materials, such as platinum, would increase the acceleration gradient to about a TeV per centimeter, about 100 times below the goal. White-dwarf or neutron-star material, which is on the order of 10,000 times denser than platinum, could give plasmas and gradients that exceed the required magnitude.

The accelerating chambers of present plasma accelerators are only several millimeters long, at best. Consequently, the amateur will probably have to run a number of machines in "tandem" in order to produce the required energy.

Based on this picture, for the UC prototype one can imagine the rotating neutron star in Cygnus X-3, which spews out 10^{6}-TeV cosmic rays, harnessed to serve as a booster for the main accelerator (though I am not sure this would be very cost-effective). A string of giant lasers stretching across the solar system would then take the particles up to the Planck energy. Admittedly, this is an unwieldy prototype, but the creative experimenter will undoubtedly find even more efficient acceleration mechanisms than the plasma accelerator and is encouraged to pursue them. Eventually, of course, one will want to go for the theoretical limit of 10^6 TeV per centimeter, at which a fully operational electron UC would be only 10^{10} centimeters long—about a fourth of the distance to the moon. A proton collider could be shorter by a factor of several million, for an accelerator length of order ten meters.

The final question remains: funding. At the projected SSC price of $250 million per TeV, the UC would cost only $\$2.5 \times 10^{24}$, something more than the U.S. budget deficit (actually about 10^{11} times the gross world product) but already a bargain considering potential spin-offs. Surely, though, one can expect that with increased technological sophistication this cost will decrease to give us the ultimate in Big Bangs for the buck. In any case, one cannot put a price tag on the philosophical benefit and change in world outlook that such a project will give to our grandchildren and our grandchildren's grandchildren, for with the help of the Ultimate Collider they will have the best chance of experimentally determining the existence of God.

POSTSCRIPT

Several readers pointed out that in the article I neglected to discuss a cooling system for the UC. The power requirements are rather large, and all the energy consumed by the accelerator must eventually go somewhere. In particular, as Richard Wolfson of Middlebury College's physics department, writing in the August 1989 *Scientific American*, mentions, the heat released by the UC would raise the surface temperature of the earth to about 12,000 °K. Indeed, as William Ossmann of Acton, Mass., put it, the "greenhouse effect" would pale in significance next to the "UC effect." To all these alert tinkerers I concede that the artist's conception may have given the impression that the UC was planned for, um, north Texas, but nowhere in the article did I dare suggest the monstrosity be built on earth.

Yet another reader, whose letter I unfortunately never saw personally, sent a $10 check to Antoni Akahito in order to be the first patron of the UC. I thank him for his farsighted generosity. At this rate, the UC should be completed in about 2.5×10^{23} years.

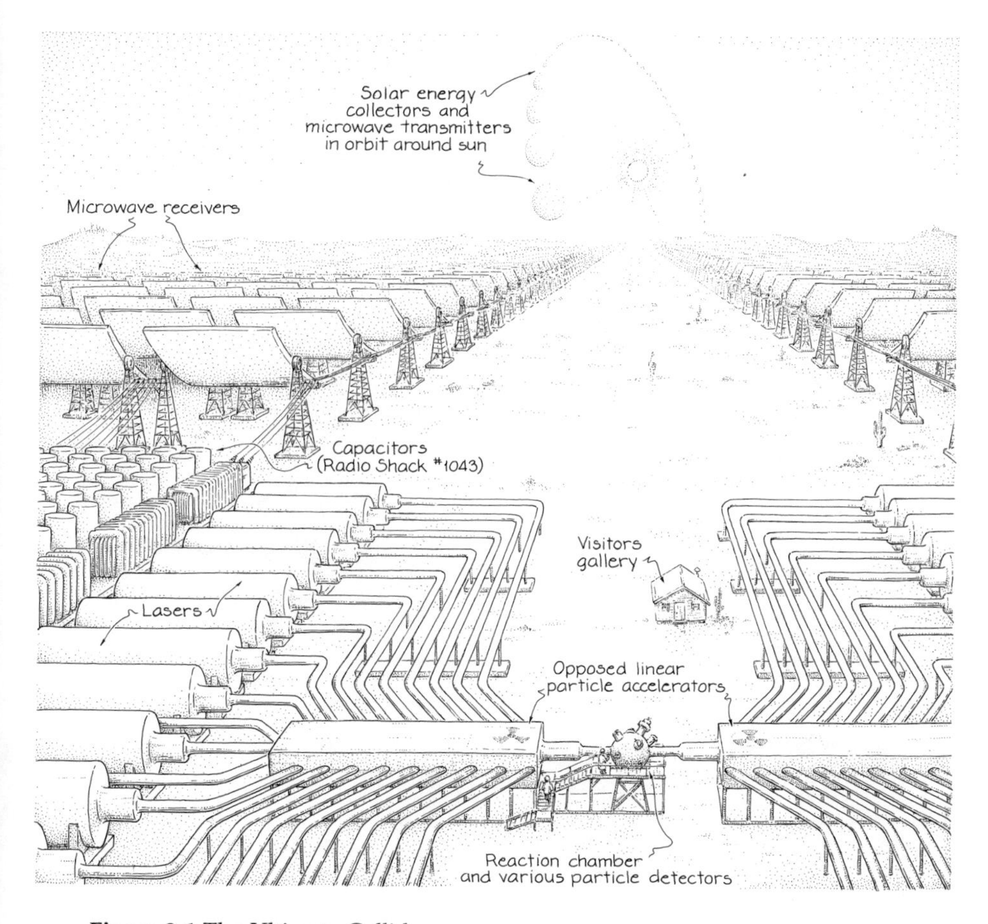

Figure 9.1 The Ultimate Collider.

Index

QMW LIBRARY
(MILE END)